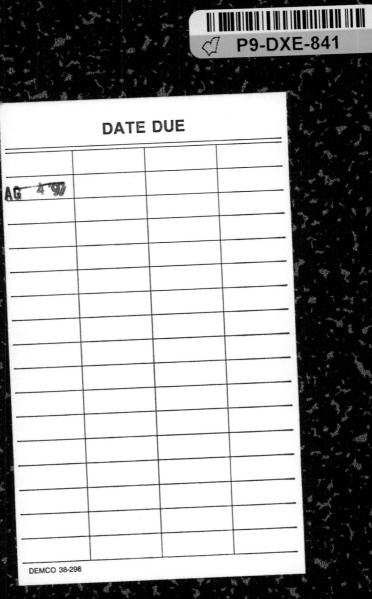

Einstein's Greatest Blunder?

Einstein's Greatest Blunder?

The Cosmological Constant and Other Fudge
Factors in the Physics of the Universe

Donald Goldsmith

Harvard University Press

Cambridge, Massachusetts
London, England

daughter, Rachel—
the music of the cosmos

————————————————————————————————————

Printed in the United States of America

Diagrams and marginal illustrations by Jon Lomberg

Library of Congress Cataloging-in-Publication Data

Goldsmith, Donald.
    Einstein's greatest blunder? : the cosmological constant and other
fudge factors in the physics of the Universe / Donald Goldsmith.
        p.   cm.
    Includes index.
    ISBN 0-674-24241-6
    1. Cosmology.   I. Title.
QB981.G594   1995
523.1—dc20   95-14762
                    CIP

SECOND PRINTING, 1995

ACKNOWLEDGMENTS

In writing this book, I have once again been fortunate to receive invaluable assistance from many astronomers, who have attempted to set me on the path of correctness. Let me regret any straying that may have occurred, while recording my gratitude to Charles Alcock, Chas Beichmann, Charles Bennett, Leo Blitz, George Blumenthal, Ken Brecher, Marc Davis, Pierre Demarque, Mark Dragovan, Richard Ellis, Jim Felten, George Field, Alex Filippenko, Wendy Freedman, Margaret Geller, Dan Gezari, Kris Gorski, Leonid Grishchuk, Paul Hodge, Craig Hogan, Bob Kirshner, Chris Kochanek, George Lake, Zoltan Levay, Andrei Linde, Phil Lubin, Dick Manchester, Steven Maran, Larry Marschall, Jeremy Mould, Tobias Owen, Bill Press, Dave Schramm, Doug Scott, Seth Shostak, Frank Shu, Joe Silk, Paul Steinhardt, Brant Tully, Mike Turner, Robert Wagoner, and Ben Zuckerman.

This project began when Howard Boyer at Harvard University Press suggested that I write a book about cosmology. It continued under Howard's successor, Michael Fisher, and received great help from Susan Wallace Boehmer, who guided the revisions and saw the book into production, and crucial assistance from Aida Donald. I am grateful to my friend Jon Lomberg for providing much of the inspiration, and all of the execution, for the diagrams in this book.

My greatest thanks go to the hundreds of astronomers, astrophysicists, cosmologists, and particle physicists who have worked so hard, and so successfully, to reveal the cosmic questions whose overview forms the subject matter of this book.

CONTENTS

Illustrations follow pages 56 and 88

ONE

Alice's Cosmic Restaurant

"The Big Bang a Big Bust?" • **"Hubble Wars Continue"** •
"The Trouble with the Universe" • **"The Cosmos in Crisis"**

In recent times, sensational headlines such as these have clamored for our attention among lesser stories of pestilence and revolt. Faithful readers of the science section of daily newspapers and weekly magazines have been confronted with the possibilities that most of the universe seems to be "missing," that during the earliest moments of time all space expanded far faster than the speed of light, and that the universe contains stars older than itself.

Any thinking layperson might reasonably wonder whether astronomers have finally overreached themselves. Can astronomers ever provide a coherent picture of the universe? If such a picture emerges, will it be anything that the average lay reader can understand? What guarantees do we have that today's picture of the cosmos will not be completely redrawn next year?

This book aims to provide some relief to those readers who fear that the familiar big-bang model of the universe may be in mortal danger. To those without that particular anxiety, I hope to provide enough basic facts about our current understanding of the universe to help them appreciate which issues in cosmology may rise to the level of a true crisis in the near future.

When a modern cosmologist—a person who studies the structure, origin, and evolution of the universe—suggests that 98 percent of the universe consists of matter unlike anything we have yet discovered, the nonscientist may feel that the mind (either his own or the cosmologist's) must reel in confusion. Yet almost every chapter of this book contains one such conclusion, and no one can hope to enjoy cosmology who remains unwilling to suspend intuitive disbelief.

Any model of the universe—unlike a model of why it rains or what makes the crops grow—creates a paradox that should have been evident to the first cosmologist who impressed his fellows around the fire with the news that the cosmos has an immense size: Neither they, nor we today, can really conceive of the universe. By definition, the universe consists of everything that exists, including volumes of space far beyond our power to observe them. A moment's reflection will prove it unlikely that our minds can wrap around the universe and thus allow us to visualize its entirety. A further moment's thought will satisfy most of us that any model of the universe which can fit into the human mind almost certainly does not include the entire cosmos. Just as we cannot expect to pass physically beyond the confines of the universe and thus to get a good look at it from the outside, so too we can hardly expect to perform the same feat in our minds. We are fated to form our ideas of the universe from the inside—a rule that governs the thoughts we are capable of having, as well as the observations we are capable of making.

Consequently, we cannot reasonably expect the universe to behave in accordance with our intuition. Formed by our limited experiences on this planet, our intuitive conclusions have no logical claim to include far-reaching truths about the universe as a whole. But being intuitive, they inescapably make such a claim. Anyone who seeks to understand and enjoy cosmology from a scientific viewpoint must continuously guard against rejecting a seemingly far-out hypothesis because it just doesn't square with our beliefs about how the universe ought to behave.

This does *not* mean that the stranger a hypothesis seems, the more likely it is to be correct. Even more erroneous than an overinsistence on intuition would be a too-easy acceptance of every fantastic supposition about the universe that comes down the pike. As in other areas of life, the trick is to withhold a conclusion based on intuition just long enough for alternative hypotheses to engage in lively competition—a competition that should yield the answer most likely to be true.

Building Models

If intuition must be held in check, what method do cosmologists use to judge competing explanations of the universe? The criteria for a good model or theory are basically fourfold:

(1) It should provide a consistent explanation of one or more aspects of what we think we know about the universe, without contradicting other conclusions that seem well assured.

(2) It must be mathematically correct, and should not violate laws of physics generally held to describe the universe.

(3) It should involve the smallest possible number of special assumptions—"fudge factors," the peculiarities that distinguish this model from others.

(4) It should offer one or more new conclusions about the universe which can—at least in theory and (far better!) in practice—allow the model to be tested. The model passes the test not simply by having its hypothesized explanation agree with what we already know, but rather (and far more convincingly) by predicting something that we do not know, but proceed to discover when we examine the conclusions that the model suggests.

To take a familiar example, the theory of gravitation first elaborated by Isaac Newton passes all four tests quite well, or it would not be so famous (see Chapter 2). It explains a tremendous amount of what goes on in the universe, and it does so with economy. We need not invoke an invisible troll beneath every vertical drop if gravity does the job. In addition, the theory of gravitation has suggested insight after insight into the universe. All of these proved correct, until Albert Einstein showed that Newton's theory must be modified. Those changes, embodied in Einstein's theory of general relativity, yielded a better theory that passed the four tests—especially the last one—with excellent ratings.

But are theoretical models truly the best way to achieve understanding? If you want to figure out why water flows downhill, the most effective procedure might be the straightforward, precise study of flowing water, not in the abstract but in the reality of a particular brook or stream. This approach, however, would miss the essence of scientific inquiry. Early philosophers quickly saw that observations of the world around them yielded a host of facts about nature—rocks are hard and water is soft; feathers float in air but raindrops fall downward—that provide the fundamental stuff of which science is made. But by themselves these facts amount to no more than what scientists call stamp collecting. Facts become useful and significant only when we begin to recognize the organizing principles behind them. A good theory provides an organizing principle that one can carry through life and apply to situation after situation. Without it, as pointed out by Charles Darwin—himself no slouch when it came to theorizing—"a man might as well go into a gravel pit and count the pebbles and describe the colors."

Every one of us creates a set of mental models, each dealing with an aspect of the world we have met, which we continually test and refine in the course of our normal lives, in much the same way that cosmologists apply the four tests outlined above. But in our daily lives, as in science, a great danger arises when we are too quick to "find" an organizing theory, and seize upon a wrong or partially wrong one rather than suspend our beliefs and admit to doubt. Parents are larger than their children, therefore large bugs must be the parents of small ones. Not so! A tossed coin comes up heads half the time; therefore if it has come up heads five times in a row, it is more likely to show tails than heads on the next toss. Not so again! (a fact that enriches professional gamblers).

Some of our mental models are simple summaries of what we see—for example, that water always flows downhill. But some are much "deeper," more insightful and flexible in application—water flows downward because gravity pulls it toward the center of Earth. This principle is a particular application of an even more general principle embodied in Newton's universal law of gravitation; it implies that on another planet water will also flow downward, but in space, far from any stars or planets, water (if it exists) may not flow at all. Newton's law of gravitation creates a model of what gravity does that allows us to understand not simply what we see on Earth but what goes on far beyond our immediate surroundings.

Of course, still greater organizing principles, still deeper models, lie beyond Newton's law. What makes gravity "work"? Why does gravity always attract and never repel? Einstein's theory of general relativity provides some of the explanation, but scientists are still seeking the complete answer to these questions—the organizing principle that not only describes the effects of gravity but also shows how and why gravity produces these effects and not others. Once we enter these realms, where we are seeking to understand the laws that govern the universe as a whole, we must accept the fact that our experience-generated intuitions count for little, and our ignorance for a great deal. A lesser species might conclude that we would do better to worry less about the cosmos and more about our next meal, but human imagination dances to a different beat.

In 1917 Albert Einstein himself confronted a "crisis in cosmology," which he dealt with in much the same way that many theorists attempt to deal with their "crises" today. Einstein realized, to his dismay, that the equations of his theory of general relativity implied either an expanding or a contracting universe. Since no one at that time suspected that the universe is indeed expanding (or, for that matter, imagined that it was contracting), Einstein introduced a new term, now called the "cosmological constant," into his equations. With the cosmological constant, mathematical models of a universe with no expansion or contraction could exist. As Einstein wrote, "That [constant] is necessary only for the purpose of making possible a quasi-static distribution of matter." Years later, after Edwin Hubble discovered that the universe is expanding (see Chapter 5), Einstein called the cosmological constant the "greatest blunder" of his life.

But today, despite repeated verification of the universal expansion and despite the evidence for the big bang provided by the discovery and measurement of the cosmic background radiation (described in Chapter 6), the cosmological constant remains alive and well in many theoretical models. Though many cosmologists believe that this constant equals zero (which is to say that it doesn't exist), others have found that a nonzero cosmological constant in the basic equations

Fudge Factors in Physics and Cosmology

first formulated by Einstein can keep their theories in agreement with observations.

The current debate over the cosmological constant provides a textbook case of how science operates. Competing theories engage in a quasi-Darwinian struggle for the minds of scientists. As they do so, some theories die forever, but many evolve into new theories, often by developing creative uses for apparently vestigial features.

In this book I have used the term "fudge factor" to describe features introduced into a theory, as Einstein's cosmological constant was, in order to resolve a pressing problem with the theory in an acceptable but aesthetically unsatisfying way. I do not mean to disparage fudge factors as foolish or unnecessary. On occasion they prove supremely useful, and even correct, so I use the term to remind the reader of the underlying motivation for creating a new feature in theorists' models of the universe. Just as we would be foolish to disdain good works by a coldhearted Scrooge, so too we should recognize that the fudge factors of the scientific world sometimes turn to gold.

Since theorists are, almost by definition, those among us with the most fertile imaginations, to bury one of their theories has proven no easier than Hercules' slaying of the many-headed Hydra. Most theories die their final death only with their originators, because the theory lies so near the heart of the theorist that he or she will find a saving modification rather than abandon it for failure to agree with observation. An outstanding theoretician like Einstein can eventually pronounce his mental child a great blunder, but even Einstein struggled for a time to "save" his theory with the cosmological constant. The chemist George Wald well summarized the problem faced by most theorists, by stating that whenever he had a new idea he made it a point immediately to stop and savor it for a while, because it would almost certainly turn out to be wrong. Only a great-souled theorist can hold this likelihood so firmly in the conscious mind.

Science's triumph has been to harness the competition among individual scientists, each committed to a personally compelling theory, to advance our overall knowledge of the universe. The current war of the cosmological models provides a fine example of this struggle, including (more than ever!) the propensity of theorists to save their theories with minor modifications.

In a 1991 article in the *Physical Review,* the cosmologist George Blumenthal summarized this flexibility by comparing one model of the universe, which contains several fudge factors whose values are not yet constrained by observational results, to Alice's Restaurant, where "you can get anything you want." For the older generation, the poignancy of this episode is that Blumenthal, who wrote his article with a much younger colleague, had to explain to his coauthor what Alice's Restaurant was all about. Still not much past thirty, his collaborator has now left cosmology for a career on Wall Street.

The Universe as We Know It

Can theorists really get whatever they want from a model of the universe? The answer must be no, or they would have no reason to debate which model best fits the actual universe. Theorists are constrained, to a greater or lesser degree, by empirical observations. Their models cannot ignore an extremely broad base of widely accepted facts which observational astronomers have painstakingly accumulated over the course of decades.

Yet it would be a mistake to assume that observations are inherently "hard" and that theories are inherently "soft." As we will see throughout the chapters that follow, many a good theory has stood its ground in the face of conflicting observational data—or no data at all. Several of the "crises" in cosmology today—such as the conflict over the age of the stars versus the age of the universe (Chapter 8)—are likely to be resolved not just by adjusting theories to fit observations but also by recognizing that some of our most rock-solid astronomical "facts" are, in fact, subject to error.

What, then, can astronomers now tell us with reasonable certainty about the composition of the universe? They would begin with *stars*—each of them a giant, glowing ball of gas, releasing energy at its center through thermonuclear fusion processes (see Chapter 3). Stars are grouped into *galaxies,* collections of billions or even trillions of stars held together by their mutual gravitational attraction (Chapter 4). A large galaxy such as our own Milky Way spans a distance of 100,000 light years—a hundred thousand times the 6 trillion miles that light travels in a year. If we model such a galaxy with a tea tray, then the sun would lie about half-way from the center to the edge, and all the stars visible to our unaided eyes on a clear night would be found within three millimeters of the sun. In this model, the sun's nine planets would all lie so close to the sun that the most powerful microscopes could barely reveal them.

Galaxies themselves form groups called *galaxy clusters,* each containing as many as several thousand member galaxies. In the tea-tray model of the Milky Way, the Local Group of galaxies—the small group to which the Milky Way belongs—would contain a few dozen tea trays, pie plates, marshmallows, and cobwebs, separated by distances that range from a few feet to a dozen yards. The closest large cluster of galaxies, called the Virgo cluster, with its center several hundred yards from the Milky Way, would span a similar distance, and would contain hundreds of tea trays, pie plates, and medicine balls, along with smaller galaxies similar to the small fry of the Local Group (see Diagram 1). As we shall see in Chapter 8, one of the great controversies of modern cosmology centers on astronomers' attempts to make this "several hundred yards" more exact—to determine the actual distance to the center of the Virgo cluster.

When astronomers look beyond the Virgo cluster, they find more great clusters of galaxies, such as the Coma cluster (a mile away in our model) and the Gemini cluster (nearly 4 miles). These "closer" galaxy clusters have actual distances of "only" a few hundred million light years (only a few hundred million times the 6 trillion miles that light travels in a year). They are named after the stellar constellations in which they appear—that is, the galaxy clusters lie in the same direction as individual nearby stars in the Milky Way which form the constellations, but at distances that are millions of times greater than the few dozen light years to those stars. Beyond these "nearby" clusters, astronomers can identify galaxy clusters at distances that are hundreds of times greater than the distance to the Virgo cluster—in our scale model, at distances of 50 miles or more.

Because these clusters are so numerous, and also so faint, astronomers have yet to map the universe in detail out to the farthest boundaries of our vision. But in recent years, they have made maps of the galaxies that lie within a few hundred million light years. Contrary to their expectations, they have found giant structures—orderings of galaxies into chains and sheets—that extend over distances as large as the map itself (see Chapter 7). Thus the question of the largest structures into which matter in the universe has formed itself remains open, awaiting mapping projects that extend to still greater distances.

In the meantime, astronomers are concerned with more pressing cosmological issues. The four currently attracting the most attention are:

(1) How did the universe manage to form such amazingly large structures during the "mere" 8 to 13 billion years since the universe began, at least in its present era (Chapter 13)?

(2) Will the expansion of space and time that started with the big bang continue forever, or will the universe someday begin to contract, perhaps toward another big bang (Chapter 9)?

(3) Of what does most of the universe consist? We know that "dark matter" exists because we can measure its gravitational effects on visible matter, but its form is

completely unknown (see Chapters 11, 12, 14). Is there enough of this missing mass to satisfy the requirements of the now-popular inflationary theory of the universe (Chapter 10), which implies a definite answer to question 2?

(4) How strong is the evidence for a nonzero cosmological constant, the "blunder" that Einstein first introduced? Assuming the existence of a nonzero cosmological constant helps some theorists to answer the first three questions, but many cosmologists are reluctant to abandon the simpler (and thus, to most of them, far more satisfying) concept of a cosmos without a cosmological constant.

This book will not answer all four questions with certainty, but it will address them in an attempt to show that observational astronomers and theoretical cosmologists, arguing back and forth between theory and observations, have not done so badly in understanding the universe. To comprehend the models of the cosmos that are currently emerging, we must be prepared to let our minds reel from time to time. Readers following the pages of this book will gain far more knowledge about current cosmology than most of their contemporaries; little enough, to be sure, in comparison with what remains to be known, but by itself a small monument to our human ability to build model empires from relatively modest amounts of information.

TWO

Gravity, Motion, and Light

From the time that the first australopithicenes looked at the starry skies and wondered what they were, humans must have attempted to interpret the complex arrangement of the lights in the heavens that seem to sail over Earth. We know from historical records that for well over two millennia most of humanity shared a common mental model of the cosmos, in which Earth forms the center of the universe. No "fact" could be more obvious—or more wrong—than the conclusion that Earth sits motionless while the sun, moon, and stars pass overhead.

The man who deserves credit for dethroning our planet from its central position was the sixteenth-century scholar Nicholas Copernicus (1473–1543). As a minor church official in what is now Poland, Copernicus had time to ponder the Earth-centered model he had studied with care during his student years in Cracow and Italy. According to this ancient Greek model, updated by the Egyptian astronomer

Ptolemy in the second century A.D., the moon, Mercury, and Venus occupy celestial spheres near Earth, surrounded by the spheres in which the sun, Mars, Jupiter, Saturn, and the fixed stars orbit. Copernicus's eventual conclusion that the sun, not Earth, ought to be the center of the cosmos was based on no observational evidence that his model provided a better fit with reality. His sun-centered cosmos drew its essence from what might be called mystical rather than scientific wells of inspiration; it was in some ways a lucky guess. Ptolemy's model required large numbers of "fudge factors" to fit with the observational data then available—but so did Copernicus's model. In fact, the latter needed even more. Later observations, however, showed that Copernicus's model far better describes the real universe than Ptolemy's did.

Mindful of the dangers of publishing unwelcome ideas, Copernicus delayed the production of his great book *On the Revolutions of the Heavenly Spheres* until 1543, when he was close to death. Though his model of a sun-centered universe impressed contemporary thinkers as an interesting revival of Eudoxus's cosmic ordering of nearly two millennia before, Copernicus's ideas made only modest headway for a century. The theory seemed to founder on the what we would now regard as the rock of observations, in particular the observations made by Tycho Brahe (1546–1601), the greatest astronomer of the second half of the sixteenth century.

Tycho's Model of the Cosmos

Tycho gained his fame from his insistence on accurate, repeated observations as the key to understanding the cosmos. Granted an island in the straits between modern

Sweden and Denmark, he created an observatory which, though it lacked telescopes (not invented until 1609), possessed instruments that allowed Tycho to measure the positions of objects in the sky with an accuracy never surpassed by other naked-eye observers. The star catalog based on Tycho's observations provided the first real advance in observational astronomy in 1400 years.

None of the stars that Tycho repeatedly observed revealed any *parallax shift,* the apparent change in stars' positions on the sky that would be caused—in Copernicus's model—by Earth's yearly motion around the sun. If Copernicus were correct, then Earth's motion in orbit, like the circle of a cosmic merry-go-round, should cause the relatively nearby stars to shift their positions against the background of more distant stars. Copernicus had already offered the correct explanation for everyone's failure to observe these parallax shifts: The stars are all so distant that their parallax shifts are too small to be measured. Tycho, a conservative in astronomical theory, offered a different interpretation that would preserve the Earth-centered model of the cosmos that he believed in deeply but would take into account the importance of the sun relative to the other planets. Tycho proposed that all planets *except* Earth indeed orbit around the sun, but the sun orbits around Earth. In other words, Tycho had a new idea for a fudge factor, one that would leave Earth in its long-cherished central position in the universe.

Tycho's model reflects a typical sort of compromise that people make during the transition from a long-held belief to a new understanding. Tycho kept as much as he could of the Ptolemaic, Earth-centered cosmos, because he felt (rightly) that a good model should not be abandoned without strong observational evidence of its invalidity. But Tycho moved part way toward the radical view of the cosmos proposed by Copernicus by making the other planets subordinate to the sun. His hybrid model with its stationary Earth was an entirely reasonable explanation for the absence of parallax shift, though it was destined to be utterly disproved (as Copernicus had foreseen) once new observational techniques allowed more accurate measurements of the stars' positions.

Kepler and the Elliptical Orbits of the Planets

During the early years of the seventeenth century, Copernicus's model of the solar system was greatly improved by Tycho's former assistant, Johannes Kepler (1571–1630), who used Tycho's many recorded observations to deduce the correct shapes of the planets' orbits around the sun.

These orbits are not the perfect circles posited by the ancient Greeks and by Copernicus but are rather ellipses—slightly elongated curves defined by the fact that from every point on the ellipse the sum of the distances to two fixed points, called foci, remains constant. If the two foci coincide, the ellipse becomes a circle; as they are set farther apart, the ellipse grows progressively more elongated, deviating ever more noticeably from circularity.

Kepler calculated that Mars, the planet most thoroughly observed by Tycho, moves in an orbit whose two foci are separated by about 5 percent of the total distance across the orbit. The sun occupies one of the two foci, while the other focus has nothing to mark it but its role in the orbital geometry. Likewise, the other planets all have elliptical orbits, with different sizes and different deviations from per-

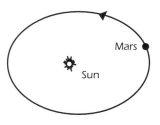

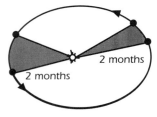

fect circularity, and with one focus point occupied by the sun. Each planet moves more slowly when farther from the sun, more rapidly when closer to it, in such a way that the imaginary line joining the planet and the sun sweeps over equal amounts of area in equal intervals of time.

Kepler's theoretical calculations impressed his contemporaries, who recognized him as a master of mathematical astronomy, but his arguments remained unconvincing to the majority of practicing astronomers. What *did* convince them that the Copernican, sun-centered model deserved credence over the Earth-centered cosmos was further observational evidence concerning the motions of the planets and their satellites, much of which was provided by the brilliant, bombastic Italian Galileo Galilei (1564–1642).

Galileo's Verification

In 1609, the same year that Kepler published his description of Mars's elliptical orbit, Galileo heard that a Dutch optician had put two lenses in line to yield magnified images of everything he saw. Quickly grasping the principle and improving on the lens design, Galileo made the first telescopes to be used for astronomical observations.

With them, he saw the rugged features of the moon—the first evidence that some celestial objects consist of much the same stuff as Earth. Galileo also observed that the planet Venus shows a cycle of changing phases, proof that we see different amounts of the sunlit half of the planet. Still more important for the acceptance of the sun-centered model, Galileo saw four satellites in orbit around Jupiter. This cut the feet from under the argument that the double motion of the moon (arising from the moon's revolution around Earth and Earth's revolution around the sun) would be too complex for nature to support. If four moons could orbit Jupiter while orbiting either the sun or Earth, surely our moon could also partake of such a double motion.

Galileo's observations, made with a primitive telescope that yielded distorted images, provoked a storm of controversy, as to both their reality and their meaning. Endowed with a native ability to take offense and to offend, Galileo more or less accidentally helped turn the Catholic Church away from its tolerant view of the Copernican model, which had arisen during the second half of the sixteenth century partly in opposition to the more "fundamentalist" Ptolemaic astronomy of the nascent Protestant movement.

Well on in years, Galileo was brought to Rome in 1630, shown the instruments that would be employed upon him if he persisted in his heresies, and thus forced into a plea-bargain agreement in which he abjured his previous statements and spent his remaining years under home arrest. The legend that at his recantation ceremony Galileo muttered under his breath, "Nevertheless, it still does move," is no more believable than any similar story about a plea bargain today. It does, however, well capture Galileo's essence: a late man of the Renaissance, sure of his own "virtu" and impatient, to put it mildly, with the bureaucratic attitudes of all government, except insofar as government could fund his research.

Isaac Newton, the man who established the sun-centered solar system as unassailably correct, was born on a farm near Grantham, England, in 1642, a few months after Galileo died

Newton and Gravity

Force of Earth on moon

equals

Force of moon on Earth

from natural causes. Educated at Cambridge and a lifelong bachelor with a mystical undercurrent in his world view, Newton was one of the inventors of the new mathematical techniques of calculus, which he used to demonstrate *why* the planets move in elliptical orbits with the sun at one focus.

The answer that Newton gave is *gravity,* the force that attracts every object with mass toward every other object with mass. Newton proposed—not just for the solar system but for the entire cosmos as well—that the amount of this force must be proportional to the product of the two objects' masses, divided by the square of the distance between their centers. This relationship became known as Newton's *universal law of gravitation.*

Thus, according to Newton, Earth attracts the sun with the same amount of force as that with which the sun attracts Earth in the opposite direction. Why, then, does Earth orbit the sun, and not the reverse? Again Newton had the explanation, contained in his famous *second law of motion.* Every object responds to the net force upon it (the result of taking into account all the various forces that act on an object) by accelerating in the direction of the net force. The amount of acceleration equals the amount of the net force divided by the object's mass. Therefore, a football accelerates far more when kicked than does a more massive stone, which is subject to greater gravitational attraction; and Earth accelerates far more in response to the sun's force of gravity upon it than the sun does in response to Earth's equal amount of gravitational force on the sun.

Using his universal law of gravitation and the second law of motion, Newton demonstrated that the natural path for a planet moving in orbit around a more massive object is an ellipse. In fact, the sun itself responds to the gravitational force from the planets by performing a tiny orbit of its own. Like the planets' orbits, the sun's orbit has at one focus not the sun but the "center of mass" of the entire solar system. However, since the sun has far more mass than any planet (a thousand times that of Jupiter, largest of the planets), this center of mass lies close to the sun's own center. Thus to a highly accurate degree of approximation we can imagine the sun to lie precisely at one focus of all the planetary orbits.

Not a publicity seeker, Newton kept his results a private treasure for many years, until his friend Edmond Halley learned what Newton had proven, persuaded him to publish, and helped shepherd through the press the most significant book yet written on physical science, the *Philosophiae Naturalis Principia Mathematica (Mathematical Principles of Natural Philosophy)*. This book, published in 1687, put the final nail through the lid of the coffin that contains the models of the universe centering on Earth. No one afterward seriously suggested that Earth does not move.

The success that Newton's laws of gravitation and motion enjoyed in explaining the motions of the heavens represents one of the triumphs of classical physics (as physicists now refer to their science as developed between the time of Newton and the early years of this century). Another achievement of classical physics—and the root of the new branch of science called *astrophysics* (the word means physics

1-pound brick

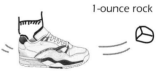

1-ounce rock

Decoding Starlight

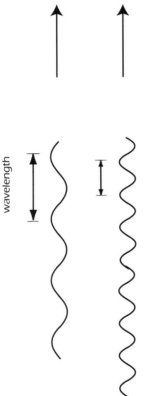

All electromagnetic radiation travels at the speed of light

wavelength

as applied to astronomy and was first coined late in the nineteenth century)—consisted of analyzing the light from celestial objects to determine their composition. To unravel the mystery of what a star trillions of miles from Earth might be made of amounts to an impressive feat of detective work, but one that astronomers have performed routinely for a century and a half.

This detective work depends on the science of *spectroscopy,* the study of the different colors of light. Newton himself was a pioneer in this field; it was he who first demonstrated that "white" light consists of light of all colors. Although the nature of light remained a subject of contention until the epoch of Einstein, well before the year 1700 Newton and his contemporaries had correctly concluded that if we regard light as a series of up-and-down vibrations—something like the waves on the surface of a pond—then we can characterize the differences among the colors of light as differences in the *frequencies* (numbers per second) of the vibrations, or alternatively as differences in their *wavelengths* (distances between successive wave crests).

Of all the light humans can see, red light has the smallest frequency and the longest wavelength, while violet has the highest frequency and shortest wavelength. Today, we know that visible light forms only a small portion of the spectrum of electromagnetic radiation, all of which is characterized by the wavelengths and frequencies of its vibrations. For all types of electromagnetic radiation, from gamma waves with the shortest wavelengths to radio waves with the longest, the product of frequency and wavelength equals a constant—the speed of light, equal to 186,000 miles per second.

Once astronomers developed instruments to spread starlight into its various colors, they found that they could profit enormously by studying this spectrum. They saw that at certain particular colors, most stars' spectra showed either a particularly large or a noticeably small amount of light, in comparison with nearby colors. These colors marked the presence either of *emission lines* (the name for frequencies and wavelengths at which much larger than average amounts of light appeared) or of *absorption lines* (for frequencies and wavelengths with a sharp reduction in the amount of light).

During the nineteenth century, astronomers realized that they could match the colors of particular emission and absorption lines with the same colors of light produced by various gases on Earth. They could then compare the spectrum of light from a particular star (our sun, for instance) with the spectra that they observed from common materials heated to the point that they glowed. If a star showed, for example, an emission line at the same wavelength as observed for a particular type of hot gas on Earth, it was reasonable to conclude that the star contained that type of gas, at least in the outer layers that produced the starlight. This experiment allowed astronomers to recognize that elements such as oxygen, nitrogen, and sodium exist in profusion in many faraway stars. Some of these elements radiate large amounts of light of certain colors; other elements block or absorb light of particular colors; and some do both at different wavelengths. In doing so, each element leaves a recognizable spectral fingerprint in the starlight.

Astronomers found that many types of atoms and molecules familiar on Earth likewise exist in the outer layers of stars. But some lines in the sun and other stars could not be

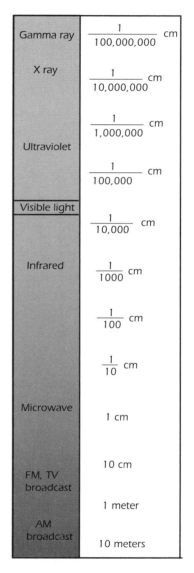

matched with any elements known on Earth. One of these unknown elements was named *helium,* after the Greek word for the sun. Helium, which turned out to be the second lightest and second most abundant element in the universe, ranks as the first element to be detected in another celestial object before being found on Earth.

From the sun to the stars, from the stars to galaxies and even beyond, the messages contained in starlight have unlocked some of the cosmic secrets only guessed at by our predecessors. Stellar fingerprints have helped to solve the mysteries of how stars produce their energy, how hot and dense they are, and how entire galaxies made of billions of stars move through space. Astronomers have identified familiar fingerprints in light that has traveled billions of light years to reach us from quasars, the most powerful and most distant individual sources of radiation in the universe, which we see not as they are but as they were billions of years ago. Our ancestors would have regarded as fantastic the notion that light not only reaches us from these enormous distances but also carries messages that can be deciphered, allowing us to map the universe and to determine the composition of its constituents. To modern readers, however, these marvelous ideas are accessible in a few pages—thanks to the hard work that astronomers have invested during the last few centuries in their quest to understand the universe.

THREE

Why Stars Shine

What makes stars shine, while planets—some of which are made of many of the same elements—produce no light of their own and can be seen only by the starlight they reflect? The answer to this key question emerged only a full two centuries after the death of Isaac Newton in 1727—as much time as had elapsed between Copernicus's first glimmering of a sun-centered cosmos and the triumph of the Newtonian theory.

Why was this question so difficult to answer? Simply because most of the cosmos differs so much from what we know on Earth. This fact so struck ancient astronomers that almost without discussion they assigned the heavens a nature utterly different from anything on Earth, a "quintessence," or fifth essence, unlike the basic four (fire, air, earth, and water) of our own world. Long centuries had to pass before terrestrial scientists discovered the elements—either rare or

25

difficult to isolate on Earth—that make up most of the stars. And still more effort had to be expended before scientists perceived the ways in which those elements interact to make stars shine. Today, astronomers think they understand the reasons for starlight, and through its analysis they have attempted to assign ages to many stars that, until recently, everyone believed to be quite accurate. But as we shall see in Chapter 8, new observations of the cosmos have called this accuracy into question.

Nuclear Fusion in Stellar Furnaces

A century ago, astronomers already knew a great deal about the composition of the outer layers of stars—the regions from which stars emit the light that we can analyze on Earth (Figure 1). By contrast, stellar interiors, the regions where stars produce the energy that makes them shine, remained a mystery, resolved only half a century ago by Hans Bethe, one of the titans of twentieth-century physics.

Stars shine because their central regions have temperatures measured in millions of degrees. In any star, energy in the form of heat (the random motion of particles) flows from the center toward the surface, from which it radiates into space. If we ask why stars are hot at their centers, the answer lies with gravity, which attracts each piece of the star toward every other piece, squeezing the entire star and thus raising its temperature—at the center most of all.

If gravity were the entire story, every star would quickly squeeze itself into the tiny volume of a *black hole,* a region with such immense gravitational force that nothing can escape from its surface. But a crucial process intervenes to prevent this disaster and to allow stars to shine for millions of years. That process is called *nuclear fusion,* the melding of two atomic nuclei to form a new nucleus.

In an atomic nucleus, protons and neutrons are held together by what physicists call the *strong* or *nuclear* force, unfamiliar to us because it acts over only extremely small distances—those characteristic of the size of a proton or neutron, about 10^{-13} centimeter. This strong force is the key to nuclear fusion. If two nuclei approach one another to distances about the size of a proton, then (and only then) the strong force is likely to make them fuse into new types of nuclei. This elemental alchemy simply does not occur at modest temperatures, such as those on or near Earth, because all nuclei repel one another through the *electromagnetic force*. Physicists have known for some time that, as we now understand the interaction of the strong and electromagnetic forces, so-called cold fusion is physically impossible, though some of them (perhaps understandably) lost their heads for a few weeks in 1989 when physicists in Utah claimed to have discovered it. Those claims were quickly—and correctly—dismissed as other scientists repeatedly failed to reproduce the critical experiments alleged to demonstrate fusion.

The electromagnetic force arises because the protons in a nucleus each carry one unit of positive electric charge (the neutrons have zero electric charge). Like charges repel one another, so at the modest velocities characteristic of modest temperatures any two nuclei headed for a collision will never approach close enough for fusion to occur. Their mutual repulsion slows them down and keeps them from coming into intimate contact.

But at high temperatures the scene changes. Then individual atomic nuclei move at enormous speeds. At temperatures close to 10 million K (10 million degrees on the absolute or Kelvin temperature scale, about 18 million degrees Fahrenheit), the nuclei reach speeds that allow some head-on collisions to bring them so close together, despite the repulsion produced by electromagnetic forces, that the nuclei fuse. This would be merely an important sidelight, rather than the secret of stellar energy, were it not for an additional fact: In most of these fusion reactions, the total mass of the particles decreases, and the total energy of motion (*kinetic energy*) increases.

From Mass to Energy

Consider, for instance, the most fundamental of all cosmic fusion reactions. At sufficiently high temperatures, a head-on collision of two protons will make them fuse. From the fusion emerge three particles: a new type of nucleus, called a *deuteron;* a *positron* or antielectron, which is the antiparticle of the electron; and a *neutrino*. The deuteron consists of a proton and a neutron bound together by the strong force. The positron will soon meet an electron, and both particles will annihilate, turning all energy locked inside them as "energy of mass" into kinetic energy, the energy of motion. The neutrino will carry its energy away from the scene of its origin, almost certain to travel for trillions of kilometers with no further interaction with any type of particle.

But if we add up the masses of the deuteron, the positron, and the neutrino (which is either zero or so close to zero that we can ignore it in this calculation), we find a total less than the mass of the two protons that collided. In

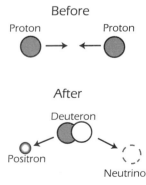

Before

Proton   Proton

After

Deuteron

Positron

Neutrino

proton–proton fusion, a small fraction of the mass (about 0.1 percent) vanishes. In its place we find additional kinetic energy among the particles that emerge from the reaction, in an amount equal to the vanished mass times the speed of light squared. Thus, Einstein's famous formula, $E = mc^2$, describes the *energy of mass* contained within any particle whose mass equals m. When mass decreases, new kinetic energy, in the amount of the lost mass times c^2, appears.

The secret of starlight resides in this new kinetic energy, made from mass that has disappeared. The neutrino carries a few percent of the new kinetic energy, while the deuteron and the positron have most of it. Through collisions with the particles around it, the deuteron will share its kinetic energy. The positron likewise shares its entire energy of mass when it meets another electron and annihilates with it, turning all its mc^2 into the kinetic energy of the photons, neutrinos, and antineutrinos (none of which has any significant mass) that emerge from the annihilation. The neutrinos and antineutrinos, like the neutrinos made directly in the proton–proton fusion, escape into space, but the kinetic energy carried by the photons is trapped and shared. In this way, the kinetic energy liberated by nuclear-fusion reactions at the center of a star diffuses outward as it is passed along by collisions among particles throughout the star.

Nuclear fusion occurs at the star's center because only there, under the enormous pressure of the star's overlying layers, can the particles in the star reach sufficiently high speeds to fuse. Gravity's invisible hand squeezes the star, making its temperature rise, most of all at its center, to the point that nuclear fusion occurs. Since the temperature

Gravitation pulls inward and heats interior of star

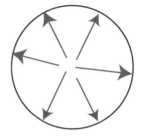

Kinetic energy released by nuclear fusion pushes outward

Star remains in balance

measures the average kinetic energy per particle, high temperatures correspond to high speeds of particle motion. Stars maintain their size because the kinetic energy released by nuclear fusion balances the tendency of the star to contract into a black hole. In that sense, nuclear fusion "saves" the star from total collapse.

From tens of millions of degrees at the star's center, the temperature falls to thousands of degrees near the star's surface. At these temperatures, the star glows brightly in the infrared, visible-light, and ultraviolet regions of the spectrum (Figures 2 and 3). Each second the glow from a mature star spreads into space the same amount of energy that the star makes at its center through nuclear fusion, though the energy takes roughly a million years to work its way from the center to the surface.

Proton–proton fusion, the basic process that produces kinetic energy in stars, leads to two more fusion reactions. First, another proton fuses with the deuteron, yielding a nucleus of helium 3 (^{3}He in scientific notation, consisting of two protons plus a neutron) and a photon. And then two nuclei of helium 3 fuse, producing two protons and a nucleus of helium 4 (^{4}He, made of two protons plus two neutrons).

Each of these steps also turns some energy of mass into kinetic energy. Since the third fusion involves two nuclei of helium 3, two each of the first two steps are needed for one of the third to occur. The net result of all three steps is that four protons fuse into one helium 4 nucleus, plus two positrons, two neutrinos, two photons, and—most important of all—additional kinetic energy.

Before

Deuteron

Proton

After

Helium 3 nucleus

Photon

The three fusion reactions that turn hydrogen nuclei (protons) into helium nuclei at the centers of stars convert about 1 percent of the total mass of the original protons into the new kinetic energy that makes the stars shine. Our sun, a fully representative star, has been fusing hydrogen nuclei into helium 4 nuclei at its center for more than 4.5 billion years. Five billion years from now, the sun will begin to exhaust the supply of protons at its center. Then, like other stars that have confronted this crisis, the sun will swell into a *red giant*, a highly rarefied ball of gas dozens of times larger than the original star; this giant ball of gas will conceal a dense core in which nuclear fusion continues.

Within its gauzy outer layer, as large as the orbit of Mercury, the core will steadily shrink, becoming ever hotter as it does so, and fusing helium into carbon nuclei. Eventually, the sun will lose its outer layers, revealing its core as a shrunken *white dwarf,* an object that no longer engages in nuclear fusion. White dwarfs consist of carbon nuclei and electrons, arranged in a crystalline lattice. They support themselves against their own gravity through a strange phenomenon called the *exclusion principle*—the absolute refusal of certain types of elementary particles to approach one another too closely.

A rare minority of stars, those that contain more than about 5 to 10 times the mass in the sun, cannot fade away quietly as white dwarfs. Instead, after their red-giant phases, these stars collapse in their centers, overwhelming the exclusion principle before it gets a chance to express itself. The collapse leads to an explosion that blows off the star's outer layers. Left behind as the collapsed core is either a *neutron*

Stellar Evolution

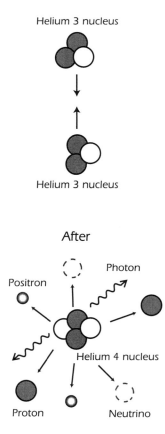

Before

Helium 3 nucleus

Helium 3 nucleus

After

Positron

Photon

Helium 4 nucleus

Proton

Neutrino

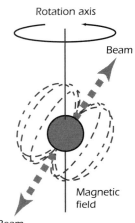

Rotation axis

Beam

Magnetic field

Beam

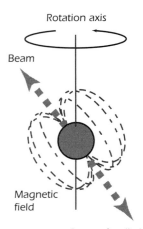

Rotation axis

Beam

Magnetic field

Beam of radiation points in different direction as pulsar rotates

star, an object only a few miles across made entirely of neutrons, or a still stranger object, a black hole, which is so condensed that its gravity prevents everything, even light, from escaping.

If the collapse produces a neutron star, that object will have an enormously strong magnetic field, because the original magnetic field in the core has been squeezed by the collapse into a much smaller volume. The collapse also affects the star's rotation: Like a figure skater who contracts her arms suddenly, the collapsing stellar core spins much more rapidly than before. The fast-spinning, highly magnetized neutron star sweeps any nearby charged particles in its web of magnetic fields, accelerating them to velocities close to the speed of light. This causes the particles to emit electromagnetic radiation—light, radio waves, and sometimes x rays and gamma rays as well. Because some regions close to the neutron star emit radiation more strongly than others, the emission from the neutron star's surroundings appears as strong pulses separated by lesser amounts of radiation. Astronomers therefore observe the object as a rotating lighthouse, a source of regular bursts of radiation called a *pulsar.*

In addition to producing a neutron star, the collapse of a massive red giant does something else, equally remarkable: It blasts the star's outer layers into space at enormous velocities, sometimes approaching the speed of light. This explosion, which masks the collapse of the star's center, turns the star temporarily into a *supernova,* which for a few months can shine with the luminosity of a billion suns.

During the final epochs before its explosion, the massive red giant fuses helium into carbon, nitrogen, and oxygen, and then fuses these nuclei to make fluorine, neon, sodium, magnesium, aluminum, and other types of nuclei all the way to iron. In its supernova outburst, all these types of nuclei explode into space, seeding the cosmos with nuclei heavier than helium. In contrast, the stars that do not explode contribute little or nothing to the general supply of nuclei, since they guard their hoard of carbon nuclei forever, fading slowly from white dwarfdom into invisibility.

Supernovae do even more than seed the universe with all the types of nuclei (save hydrogen and helium) up to iron (number 26 in the periodic table). In addition, the fury of the supernova explosion itself creates all the nuclei *heavier* than iron. During the explosion, iron nuclei fuse, producing nuclei such as copper and zinc; krypton, strontium, and molybdenum; silver, tin, lead, gold, mercury, and uranium. Because supernovae have only a brief moment to fuse nuclei heavier than iron, these nuclei are much rarer in the cosmos than iron and the lighter elements.

In short, when it comes to making the heavier elements, supernovae basically do it all. We owe to supernova explosions the fact that we find any elements other than hydrogen and helium (themselves made soon after the big bang) on Earth or in the heavens. Every molecule in our body contains some of these star-made elements: carbon, nitrogen, oxygen, phosphorus, or other nuclei made in long-vanished stars that shot those particles into interstellar space as they died.

Variable Stars

Most stars, like our sun, shine steadily for billions of years, slowly fusing hydrogen into helium at a constant rate. As a star ages, however, the supply of hydrogen nuclei (protons) in its core becomes exhausted. Stars adjust to this situation as best they can, subject to the basic rule that their own gravitation tends to squeeze them, while their release of energy through nuclear fusion opposes this tendency. As a star's core ages, it contracts, squeezed by the gravitationally induced pressure, and steadily raises its temperature. Higher temperatures allow additional nuclear-fusion reactions to occur, so star manages to keep shining, but the star's adjustments to the changing situation can produce irregularities in its output.

Toward the end of their proton-fusing phases, many stars become *variable,* fluctuating in their luminosities. As the sun ends its red-giant phase, it will spend many million years as a variable star. This stellar variability sometimes consists of rather random fluctuations, but more often appears as regular, up-and-down variations, producing what astronomers call a *periodic variable.*

The best known of the periodic variable stars are the subclass called *Cepheid variable stars,* named after their first representative to be studied, the star Delta Cephei. Cepheid variables have the virtue, astronomically speaking, of possessing great intrinsic luminosities. This allows astronomers to find and to study them even at relatively great distances. But Cepheids have an even more useful property, discovered early in this century by Henrietta Swan Leavitt through observations of these variables in a collection of stars called the Small Magellanic Cloud.

During the first voyage that circumnavigated Earth, Ferdinand Magellan and his sailors saw two "clouds" that appeared to move with the stars. We now know that these objects, called the Large Magellanic Cloud and the Small Magellanic Cloud, are satellite galaxies of our own Milky Way, and the closest galaxies to our own. The Magellanic Clouds lie so close to the south pole of the sky that they never rise above the horizon of observers farther north than Puerto Rico.

Working from photographic plates at the Harvard College Observatory, Leavitt—assuming (quite correctly, so far as we can tell) that all the stars in the Small Magellanic Cloud have about the same distance from us—found that any two Cepheids with the same period of light variation also had the same average luminosity; that is, they emit the same amount of energy per second, averaged over one of their cycles of variation. Cepheids with longer periods have greater luminosities than those with shorter periods. If astronomers can identify two Cepheid variable stars with the same period of light variation, the one that appears fainter to us must be farther away. Indeed, since the apparent brightness of any object decreases in proportion to the square of its distance from an observer, if one Cepheid variable has four times the apparent brightness of another with the same period, the fainter one must be twice as distant.

Because Cepheids have intrinsically large luminosities, they can be discovered at relatively great distances, making them highly useful "standard candles" for astronomical research. As we shall see in the chapters that follow, Cepheid variables underlay the first determinations of the distances to galaxies, and they lie at the forefront of today's discussions about galaxies' distances and the rate at which the universe is expanding.

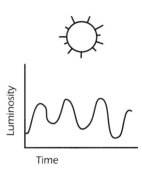

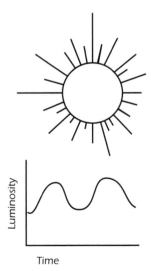

FOUR

Mapping the Milky Way

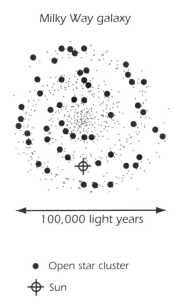

Milky Way galaxy

100,000 light years

● Open star cluster

⊕ Sun

In Newton's era, astronomers knew quite well that the sun is just one star among many, but they had no good answer to the question of how stars are distributed throughout the universe. Do stars sprinkle the cosmos at random? Or are they concentrated at certain places in preference to others?

A part of the answer appears almost immediately to anyone who surveys the sky with a small telescope: Stars do cluster together in certain regions. The most obvious clumping is the band of light that became known as the Milky Way. Galileo's crude telescopes revealed that the Milky Way arises from countless individual stars concentrated in space, but three centuries had to pass before astronomers fully understood this agglomeration. Today we recognize that the band of stars we see in the night sky is part of a flat, disklike assemblage of stars, called a *galaxy*, of which our

36

solar system is a member (Diagram 2). When we look outward from the plane of the disk, we see relatively few stars,
but when we look in directions around the sky that lie along
the disk, we observe the glow of our own Milky Way galaxy.

The second type of stellar concentration visible in a
small telescope is called a *star cluster*. Star clusters come in
two types, *open clusters* and *globular clusters*. The nearest
open clusters are the Hyades and the Pleiades, both located
in the constellation Taurus, though the Pleiades are about
three times more distant than the Hyades. Like all open clusters, the Hyades and Pleiades each contain a few hundred
stars, which astronomers feel sure all formed together and
are only slowly drifting apart. With a (good) unaided eye,
you can see the seven brightest members of the Pleiades,
affectionately named the Seven Sisters—a stylized picture of
which, engraved on the medallion of Subaru automobiles,
shows the appeal of this star group in cultures besides our
own.

Globular clusters represent mightier concentrations
than open clusters do; they each contain several hundred
thousand, or even a few million, stars. The best known of
the globular clusters lies in the constellation Hercules, visible
in summertime to anyone with good eyes or a pair of bin-
oculars. Like this cluster, globular star clusters contain sufficient mass to hold themselves together indefinitely, as each
of their stars orbits the cluster's center of mass. Unlike open
star clusters, globular clusters have turned all their original
gas and dust into stars, so they no longer engage in star
formation, and have not done so for billions of years. Open
clusters, by contrast, are often much younger than the glob-

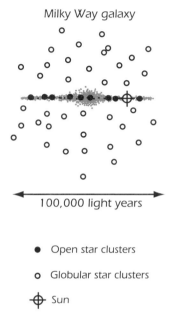

Milky Way galaxy

100,000 light years

● Open star clusters

○ Globular star clusters

⊕ Sun

ular clusters and contain large gas clouds that continue to give birth to new stars.

How We Know the Ages of Star Clusters

Star clusters provide far more than a chance to study large groups of stars, apparently born together from giant clouds of gas and dust. By comparing the observed characteristics of stars in a cluster, astronomers can estimate the ages of these stars: the amount of time since they began to fuse hydrogen into helium at their centers. The technique of dating star clusters began 40 years ago, as astronomers came to understand what makes stars shine, and has been steadily refined through ever-better computer models of what goes on in stellar interiors.

What is that technique, and why does it apply only to star clusters? The technique involves the *temperature-luminosity diagram,* a graph of the surface temperatures of stars (plotted on the horizontal axis) versus the stars' luminosities (plotted vertically). Through an oddity of astronomical history, the horizontal temperature scale in such a diagram has temperature increasing to the *left!* Only a strong tradition could overcome logic to this extent, leaving a sizable danger that those less steeped in tradition will misread the temperature-luminosity diagrams that astronomers use so often and so well.

In an ideal situation, astronomers know the distances to stars in the cluster, and thus can calculate the stars' luminosities, also called their *intrinsic (absolute)* brightnesses. But if the cluster's distance cannot be determined, astronomers can work with the stars' *apparent* brightnesses. Since all the

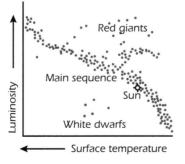

Temperature-Luminosity Diagram

Red giants

Main sequence

Sun

White dwarfs

Luminosity

← Surface temperature

stars have almost the same distance from us, their apparent brightnesses have the same ratios as their absolute brightnesses, and the comparison of absolute brightnesses reveals much about the ages of the stars in the cluster.

When astronomers examine the stars' surface temperatures and brightnesses, they find that the points representing most of the stars in a cluster fall in a particular region of the diagram, called the *main sequence,* a swath that extends from stars of low surface temperature and low luminosity at the lower right corner of the diagram up to stars of high surface temperature and high luminosity at the top left corner. Our sun is a main-sequence star, placed just about in the middle of the main sequence by its surface temperature of 5,800 degrees absolute and its luminosity of 4 × 10^{33} ergs per second. (We know the age of the sun, 4.5 billion years, not by studying the sun itself but by dating the oldest rocks found on Earth and rocks returned from the moon, as well as from our calculations of how the solar system formed. Those calculations include the assumption—entirely justified, in most scientists' opinion—that the sun's family of planets formed along with the star itself.)

In addition to its main-sequence stars, star clusters reveal some white dwarfs, whose low luminosities but relatively high temperatures place them *below* the main sequence (lower center of Diagram 4), and some red giants, with high luminosities and relatively low temperatures that place them *above* the main sequence (at the upper right corner of the diagram). The red giants are former main-sequence stars that have now begun to cool as they exhaust the supply of protons for nuclear fusion at their centers. The white dwarfs have evolved past the red-giant stage and have lost their outer layers to reveal their hot, shrunken, naked cores.

How can a cluster of stars, all born at the same time, contain individual stars that occupy such different points along their career paths? Here lies the secret of dating star clusters. Calculations show that more massive stars pass through their main-sequence phases more rapidly than do low-mass stars. More massive main-sequence stars have higher surface temperatures and larger luminosities, which place them at the top of the main sequence on the temperature-luminosity diagram. As the stars grow older, they start to feel the pangs arising from the exhaustion of protons in their cores. They then expand and cool, which moves their points to the right on the temperature-luminosity diagram, toward lower surface temperatures.

The first stars to leave the main sequence will be the most massive ones in the cluster, followed by the stars with slightly less mass. Then, in a steady progression, stars will peel from the main-sequence toward the red-giant region of the temperature-luminosity diagram in order of decreasing mass. When astronomers examine the diagram of a given cluster's stars, they find that the main sequence extends up to a particular surface temperature and luminosity, but no farther. The stars on the main sequence with the highest surface temperatures and luminosities are just about to become red giants. Hence by measuring the luminosities and surface temperatures of the stars at the top of a cluster's main sequence, astronomers can determine the age of all the stars in the cluster—if they have accurate models of how stars grow old.

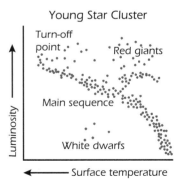

Young Star Cluster

For years, astrophysicists have worked on their models of stellar evolution, taking into account each star's mass as well as additional parameters, such as the fraction of the star that consists of elements other than hydrogen and helium, which may affect the star's evolution. Their models allow them to assign ages to those star clusters whose temperature-luminosity diagrams have been plotted from observations of hundreds of individual stars.

The ages of open clusters, derived by locating the top end of the main sequence, range from a few million to many billion years. In contrast, *all* globular clusters have ages of many billion years; the Milky Way and other galaxies stopped making globulars long ago. We see this in the fact that the main sequences of these clusters terminate at relatively low surface temperatures and luminosities, close to those of our own sun, whose total lifetime as a main-sequence star will be about 10 billion years. Using this technique, astronomers have assigned an age of 10 billion years to the globular cluster M13, and even higher ages to a few other globular clusters. At the top of the age range, we find the cluster Omega Centauri, 15,000 light years away, one of the most striking sights in the skies above the southern hemisphere. By constructing the temperature-luminosity diagram for the stars in Omega Centauri, astronomers have found that this globular cluster's stars are nearly 16 billion years old.

Or are they? In Chapter 8, we shall examine the evidence that the universe has an age billions of years *less* than 16 billion years—possibly just *half* the calculated age of Omega Centauri—and consider what to do about this con-

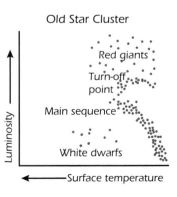

tradition. But before we approach the conundrum of a universe younger than its parts, we must complete our survey of the types of objects that astronomers have spotted in the cosmos, in particular of the fuzzy clouds that astronomers have called nebulae.

Cataloguing the Cosmos

As astronomers improved their telescopes during the eighteenth and nineteenth centuries, they discovered hundreds of star clusters, all of which could be classified as either open or globular. To astronomers of the early twentieth century, the arrangement of matter in the heavens seemed fairly well established. The Milky Way seemed to include all the stars, star clusters, and gas clouds that telescopes revealed. But one unanswered question, destined to change the entire model, was prominent: the nature of the "spiral nebulae."

A small fraction of the objects that glow in our night skies exhibit not the pointlike appearance of stars but rather a fuzzy shape. Much of the advance of our knowledge of the structure of the universe in the twentieth century derives from our increased understanding of these fuzzy objects, generically called *nebulae,* from the Latin word for "cloud."

The most prominent nebulae were catalogued during the late eighteenth century by Charles Messier, a French amateur astronomer who was searching for comets and made a list of noncometary fuzzy objects that might confuse him in his efforts. His list of "Messier objects," numbering just over a hundred, has carried Messier's name through the centuries, though today his comet discoveries are forgotten.

As might be expected, Messier's list offers a mixed bag

containing the brightest nonpointlike objects visible in the night skies of France. A few of the objects later turned out to be star clusters, either open or globular, which Messier would have recognized as such had he possessed a better telescope. A few others are *planetary nebulae,* confusingly named from their resemblance to a planet's disk when seen in a small telescope. In fact, planetary nebulae are expanding shells of gas ejected from aging stars, heated from within and made to glow by the star's radiation.

Many other Messier objects were eventually recognized to be huge masses of hot, glowing interstellar gas heated by young stars buried within. Like all stars, these groups of young stars had formed through gravitational contraction of the gas, and eventually many more stars could be expected to condense within the gas cloud, initiating nuclear fusion at their centers under the influence of gravity. Spectroscopy revealed that these hot gas clouds, which were then called "diffuse nebulae" and which today we call *HII regions,* were primarily made up of hydrogen, helium, carbon, nitrogen, oxygen, and neon. The term HII, meaning ionized hydrogen, refers to hydrogen atoms that have lost their electrons (thus becoming hydrogen "ions") under the influence of intense radiation from nearby stars.

The closest large HII region to us, the Orion Nebula, forms the middle "star" in Orion's sword, but in fact consists of an open cluster of stars seen early in its formation process (Figure 2). The young hot stars that have already formed illuminate and heat the gaseous cocoon around them. Spectroscopic analysis of the light from this gas reveals a temperature close to 10,000 K. Other HII regions, similar in size and nature to the Orion Nebula, are dotted among the stars of the Milky Way. All are basically the same type of object, an open star cluster in its early formation stages. The Orion Nebula has already produced a few hundred stars and will eventually give birth to many thousand.

But Messier had catalogued another type of nebula, one that appeared to consist of webs of gas twisted into complex spiral patterns. These "spiral nebulae" seemed somewhat akin to the diffuse nebulae and the planetary nebulae, but no one could say for sure. For half a century and more, astronomers drew the elaborately structured spirals they saw through their best telescopes, and by the end of the nineteenth century good photographs complemented these drawings. But their precise nature had not yet been determined.

By the early twentieth century, a minority of astronomers proposed that the spiral nebulae are not clouds of gas at all but rather collections of stars so distant that we cannot see their individual members. If this is so, the spiral nebulae must have truly immense distances from the solar system, distances that would place them beyond all the stars and star clusters that astronomers had identified. As telescopes improved, astronomers saw that the spiral nebulae are indeed giant collections of stars, with some hot gas clouds and dark dust particles strewn among them.

But how large, and how distant? Were the spiral nebulae merely enormous star clusters in our own Milky Way, or were they so distant that they must equal the Milky Way in size, and thus be "island universes" in their own right? Does the group of stars that we call the Milky Way include the entire visible universe, as astronomers of the nineteenth century believed, or is it just one structure in a cosmic sea of many such structures?

Edwin Hubble and the Nature of Spiral Nebulae

The man who demonstrated the true arrangement of stars in the universe was Edwin Hubble. Born in 1889 in Missouri, Hubble was a Rhodes Scholar who had nearly become a lawyer but instead opted for astronomy. He was also a veteran of overseas duty in World War I who loved to be addressed as "Major," a country boy who affected an English accent, and a brilliant astronomical observer who seemed to sense which problems would prove most fruitful for study. During the years after World War I, Hubble had access to the recently completed 100-inch reflecting telescope on Mount Wilson, California, then the world's largest, built on a mountaintop overlooking the Los Angeles Basin (a site then free of smog from the world's largest agglomeration of automobiles). Hubble decided to use his observing time to resolve the nature of the spiral nebulae.

The first quarter of this century brought a discussion, later characterized as a "Great Debate," among astronomers on this issue. Harlow Shapley, the young director of the Harvard College Observatory, had made an estimate of the size of the Milky Way based on a map of the globular star clusters that it contains. According to Shapley, this size was so great—about three times our present measurements—that the Milky Way could contain even the spiral nebulae whose distances could only be guessed, and which would then be subunits of the Milky Way. Other astronomers, most notably

Heber Curtis of the Lick Observatory in northern California, asserted that Shapley had overestimated the size of the Milky Way and that the distances to spiral nebulae were far greater than had been previously assumed. What astronomers needed was a direct determination of the distance to one or more of the spiral nebulae, and this Hubble provided.

Hubble used the 100-inch telescope to search for Cepheid variable stars in the brightest spiral nebulae. As Hubble realized, Henrietta Leavitt's discovery of the period–luminosity relation in these stars meant that if two Cepheids with the same period of variation can be identified, the one that appears fainter must be farther away from us. Before long, Hubble had identified Cepheids in several well-known nebulae, including the great spiral in Andromeda. He compared the apparent brightnesses of these stars with similar but brighter Cepheids in the Milky Way (whose comparatively short distances from us had been estimated by other means) and showed that the Andromeda spiral nebula must have a distance of at least half a million light years. Since light travels at 186,000 miles per second, one light year— the distance light covers in a year—equals about 6 trillion miles. No astronomer interested in cosmology would use a unit of distance so tiny as a mile, and, as Hubble's observations showed, even a light year proves utterly modest in describing cosmological distances.

Half a million light years was too large for the Milky Way, whose size had been estimated by Shapley at no more than 300,000 light years. Modern revisions of the work done by Hubble and others assign a distance of 2.2 million light years to Andromeda, and a diameter of about 100,000 light years to our Milky Way. Thus the Andromeda galaxy lies at a distance twenty times the Milky Way's diameter. It therefore must be an independent system of stars.

By using the word "galaxy" (derived from the Greek word for "milk," and honoring our own Milky Way as the prototype), we enter the modern era of cosmology, in which the spiral nebulae have been recognized as *spiral galaxies.* These flattened, disklike collections of gas, dust, and billions of individual stars have prominent spiral arms within the disk, where the youngest, hottest stars reside and where new stars are forming (Figure 4). Galaxy catalogs now list thousands upon thousands of galaxies, a sizable fraction of which are spirals like our Milky Way (Figure 5).

An equally sizable fraction of all known galaxies fall into the category of *elliptical galaxies.* They have about the same diameters as the spirals, show little if any flattening, and have little or no interstellar gas and dust. Apparently the ellipticals had more efficient star-formation processes than the spirals. Unlike spiral galaxies, they have not formed any new stars for billions of years.

The remaining minority of galaxies are *irregular galaxies,* typically much more formless and somewhat smaller than the largest spirals and ellipticals. The closest and best-studied examples of irregular galaxies, the two Magellanic Clouds, are satellites of the Milky Way. Their masses each contain a few percent of the mass in our galaxy but, like most irregulars, are especially rich in the interstellar gas and dust that continues to form new stars (see Figure 6).

So far as astronomers can tell, nearly all galaxies began to form within a few billion years after the universe began.

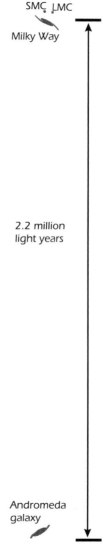

SMC LMC

Milky Way

2.2 million
light years

Andromeda
galaxy

What made these galaxies form? And why do they exhibit these different patterns? These questions remain unanswered; indeed, as we shall see in Chapter 13, they lie at the heart of modern cosmological research.

Today the essential questions about the structure of the universe no longer concern the distribution within galaxies of star clusters, gas clouds, and dust. Instead, astronomers investigate the arrangement and motions of entire galaxies and especially *groups* of galaxies throughout the universe, at immense distances in space and time from the Milky Way. These observations, and the conundrums they produce, will dominate the rest of our survey of cosmology.

FIVE

The Discovery of Universal Expansion

Edwin Hubble's announcement that spiral galaxies lie far beyond the confines of the Milky Way came in 1923, a year marked for most Americans by the death of President Harding at the Palace Hotel in San Francisco. The term-and-a-half of the next president, Calvin Coolidge, saw the United States pass through a frenzy of expanding business activity, which peaked with the great stock market crash that brought the 1920s to a close. By a sort of cosmic coincidence, the year of the crash, 1929, was also the year when Hubble's efforts culminated in the discovery that made him the most famous astronomer of the first half of this century: the expansion of the universe.

Hubble spent the years from 1923 to 1929 making ever-better estimates of the distances to galaxies, relying on Cepheid variable stars as his basic measuring device. He became interested in whether the *distances* that he estimated were correlated with the *motions* of these galaxies. These motions had been investigated by other astronomers, who studied the spectra of the light from galaxies—light that astronomers now recognized as a melange from billions of stars. In this mixture, astronomers could recognize some of the most prominent features they had found in the spectra of individual stars. For example, they saw the dark absorption lines arising from the blockage of light by sodium and calcium ions, which rank among the most prominent features in the spectrum of the sun and many other stars. But they also saw that the wavelengths of these absorption lines differed slightly from those observed in a terrestrial laboratory. The difference almost certainly arose from a cause familiar to all scientists, the Doppler effect.

The Doppler Effect

As the early-nineteenth-century scientist Christian Johann Doppler had perceived, the motion of a source of waves of any kind—sound waves, water waves, or light waves—will make the waves look slightly different to an observer than they would if the source were not moving. The term *Doppler effect* describes the fact that if the source of waves approaches the observer, each successive wave crest will arrive a bit sooner. Conversely, if the source recedes from the observer, the wave crests will each arrive a bit later than they would if no motion occurs. In other words, the measured wavelength—the distance between successive wave crests—will be less for a source approaching the observer, and greater for a source receding from the observer.

The Doppler effect also changes the frequency—the

time interval between the arrival of successive wave crests. Motion toward the observer increases the measured frequency, whereas motion away from the observer decreases it. The amount of the change in wavelength and frequency is called the *Doppler shift* and arises directly from the Doppler effect. More rapid motion produces a larger Doppler shift. Furthermore, the Doppler effect does not depend on whether the observer moves instead of the source—or whether they both move. Whatever does the moving, astronomers know that if they can measure the change in wavelength arising from the Doppler effect, they can determine the amount of relative motion between the source and the observer.

Now the frequency of any wave, multiplied by its wavelength, will always equal the speed at which the wave travels. Therefore, since the speed of light remains constant (a fact verified by repeated measurements during the late nineteenth and early twentieth centuries), a larger frequency for light waves implies a smaller wavelength, and vice versa. Astronomers often call a shift to lower frequencies and longer wavelengths a *red shift,* because red light has the lowest frequencies and longest wavelengths of all the colors of visible light. They use the term *blue shift* for a Doppler shift to shorter wavelengths and higher frequencies, even though they might more properly call this a violet shift. But blue will do, and has done for years.

The Doppler effect offers us a chance to determine how rapidly an object that emits light is approaching us or receding from us—if we can determine the amount of the Doppler shift in its light. But how can we determine the

In a source moving toward observer, the waves arrive in a compressed manner, that is, with shorter wavelengths

In a source moving
away from observer,
the waves appear stretched
to longer wavelengths

change produced by the Doppler effect if we measure only the *observed* frequency and wavelength, and not their values in the absence of relative motion? The answer, as astronomers had come to recognize by the early years of this century, is that the spectra of stars show familiar patterns; though they exhibit many variations, the spectra possess an underlying similarity.

Since the absorption lines produced by sodium and calcium were already familiar, when astronomers saw strong absorption lines at slightly different wavelengths from the usual ones, they could recognize them as their old friends, displaced by the Doppler effect. If a spectrum contained only one or two absorption lines, this might seem a bit of a stretch, but improved observations revealed many more spectral lines. Galaxies' spectra became as familiar as a child's face—adjusted by the results of the galaxies' motions with respect to those who observed them. By now, astronomers who study galaxies' spectra can recognize familiar patterns in the ratios of wavelengths of even two or three of the most prominent spectral lines.

Hubble's Law

During the early years of this century, Vesto Slipher, an astronomer at the Lowell Observatory near Flagstaff, Arizona, measured the Doppler shifts of several dozen galaxies. Even this small sample, increased by a few other measurements at other observatories, revealed a striking nonrandomness. Only a very few galaxies had relative motions of approach, while by far the majority of galaxies were receding from the Milky Way. This was suggestive by itself (though in that more modest era astronomers refrained from speculating about the discrepancy). It was left to Edwin Hubble to find the explanation for the lopsided distribution of the galaxies' Doppler shifts.

Hubble's work allowed him to pair each galaxy's Doppler shift with the distance that he estimated for that galaxy. When he made a graph of the outcomes, he saw a definite trend: Only a few of the closest galaxies are approaching us, while the majority of galaxies are receding, with speeds that increase in proportion to their distances from us.

Hubble could see what this means. The motions of galaxies that we observe have a general pattern which matches progressively larger distances with progressively greater speeds of recession. But each galaxy also has a modest motion of its own, in a random direction, like a flea hopping on the back of a running dog. Because these individual motions are modest, only among the closest galaxies do they have a chance to counteract completely the tendency to recede, and thus to produce a velocity of approach. For more distant galaxies, the galaxy's individual motion merely adds slightly to, or subtracts slightly from, the galaxy's much larger velocity of recession.

Hubble's original graph, published in the *Proceedings of the National Academy of Sciences,* presented the data from only a few dozen galaxies. Nevertheless, Hubble drew the conclusion—soon supported by much more observational evidence—that, except for the closest galaxies, all the galaxies that we observe obey what astronomers now call *Hubble's law:* Galaxies are receding from us with speeds proportional to their distances. (Today we can see that Hubble's

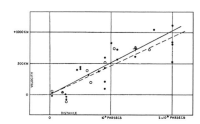

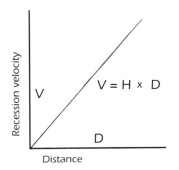

first diagram does not really show the overall behavior of galaxies, so his conclusion was a fortunate prolepsis, that is, a good guess.)

In algebraic notation we write Hubble's law as $V = H \times D$, where V is the galaxy's recession velocity, D is its distance, and H is a constant throughout the universe, which we call Hubble's constant. We shall see in Chapter 9 that this statement is true at any given time, but that H, the constant of proportionality between distance and velocity, may change slowly over time.

Within a few years, Hubble and his collaborator Milton Humason had extended their distance estimates outward into space by a factor of twenty. They continued to find a straight-line relationship between galaxies' recession velocities and their distances from us, but now they were dealing with entire *clusters* of galaxies, which appeared to be the basic structural unit of the universe. This work established Hubble's law to the satisfaction of all but a very few doubting astronomers, and later work has continued to confirm Hubble's original conclusion that galaxies' recession velocities increase in direct proportion to the galaxies' distances from us.

Can we explain Hubble's law without making the Milky Way the center of the universe? Hadn't Hubble inadvertently restored the preeminent position from which Copernicus and Heber Curtis had so brilliantly winkled us, like clams displaced from their happy homes in the sand? The answer is no—unless we are so determined to make our position central that we abandon an entirely reasonable assumption.

That assumption is called the *cosmological principle,* and states that our view of the universe is representative of the entire universe at this time. If we accept the cosmological principle as a good working hypothesis, then observers in every galaxy in the universe should see the same phenomenon that we do: galaxies receding in all directions, with speeds proportional to their distances *from that observer.* In that case, the entire universe must be in a state of expansion, since galaxies are moving away from one another, everywhere.

And how can this be so? Intuition implies that you can't have objects everywhere receding, and nowhere approaching. In fact this is quite possible—if we suspend, at least for the time being, our difficulties with the "edge of the universe." Imagine a balloon with spots on its surface. Blow up the balloon, and every spot moves away from every other. Sit on any spot, and look around the surface of the expanding balloon; you will see spots receding from you, at speeds proportional to their distances.

So Hubble's law indeed holds true on the surface of an expanding balloon. All that remains is to use the expanding balloon as a model for the actual universe! But this feat presents most of us with a problem that cannot be entirely explained away. The universe is *not* a balloon, though a balloon can help to model it. *If* we can imagine all of actual three-dimensional space reduced to the two-dimensional surface of a balloon, and *if* we can likewise imagine that light travels only around the balloon's surface (for nothing else exists), *then* we can perhaps believe that the entire universe

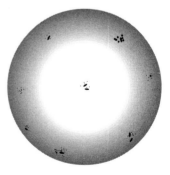

can exist in a state of expansion. What must we do with the universe inside and outside the balloon? Simply imagine it out of existence—if we can.

Balloon or no balloon, Hubble's results galvanized cosmologists. The cosmological principle seemed completely reasonable, and the expanding universe therefore seemed the most natural interpretation of Hubble's work.

In addition, modern theories of physics added support to what Hubble had found. In 1916 Albert Einstein had published his general theory of relativity, which described in a few mathematical equations how matter bends space. Gravity curves space more strongly in regions that lie closer to large amounts of matter. Alexander Friedman, a Russian mathematician, recognized a few years afterward that Einstein's equations imply that the universe must always be in a state of either expansion or contraction. In other words, in their purest form, Einstein's concepts about space, time, and matter made it inevitable that space in the universe cannot be static. Our most fundamental concept about space—that space simply "sits there"—is, according to this early work of Einstein and Friedman, just plain wrong. (In fact, a much deeper objection to space just "sitting" is that the notion implies that space has an existence independent of matter— a notion that many cosmologists would dispute.)

Within a few years after Hubble's first publication in 1929, the expanding universe was accepted as fact throughout the astronomical world, though a handful of dissenters remain to this day. Everything that astronomers can see exists in a state of expansion—not from any one central point, but from everywhere, the fleeing of all from all.

Figure 1. The open star cluster Messier 16. Legend follows.

Figure 2. The Orion Nebula. Legend opposite.

Figure 3. The Horsehead Nebula. Legend below.

Figure 1. Messier 16 is a young open star cluster located in one of the Milky Way's spiral arms, about 6,500 light years from the solar system. This cluster formed a few million years ago from great clouds of gas and dust that are still giving birth to more stars. **Figure 2 (opposite).** The Orion Nebula is the closest large star-forming region to the solar system. This young open cluster, which contains dozens of young hot blue stars, forms the middle "star" in the sword of the constellation Orion. **Figure 3 (above).** Close to Orion's belt lies the Horsehead Nebula. This dark mass of dust, which absorbs starlight of all colors, produces the shape of a horse's head against the background of scattered light from a highly luminous star beyond the edge of the photograph. Photographs 1–3 by David Malin; photograph 1 © Anglo-Australian Telescope Board, 1980; photographs 2 and 3 © 1979 Royal Observatory, Edinburgh, Anglo-Australian Telescope Board

Figure 4. The spiral galaxy Messier 83. Legend opposite.

Figure 5. During 1990 and 1991, the Cosmic Background Explorer (COBE) satellite photographed the Milky Way in four different infrared frequency bands. From these images, the COBE astronomers produced this photograph of the parts of the Milky Way closer to the galactic center than our solar system. Because we are well toward the outside of the galactic disk, this image makes the Milky Way look much like other spiral galaxies, if we happen to see them edge-on. Courtesy of NASA/Goddard Space Flight Center

Figure 4 (opposite). About 10 million light years from the Milky Way, the spiral galaxy Messier 83 presents a nearly face-on appearance. This galaxy resembles our own in its size and structure, as well as in the concentration of yellowish stars near its center and of young hot, luminous stars in its spiral arms. Numerous supernova explosions have been observed in Messier 83, at an average rate of one every two decades. Photograph by David Malin, © Anglo-Australian Telescope Board, 1977, 1979, 1980, 1981, 1982, 1986

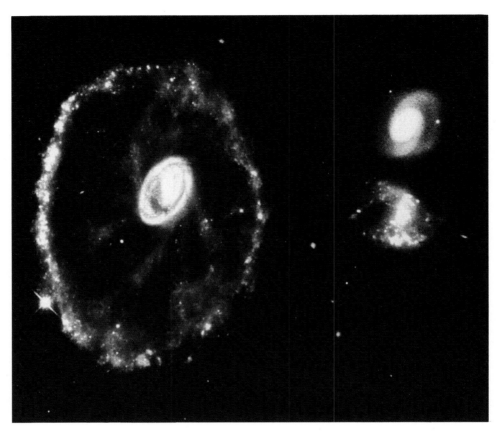

Figure 6. In October 1994 the Hubble Space Telescope obtained this stunning image of the Cartwheel Galaxy (about 500 million light years away), through which one of the two small galaxies to the right has recently passed (we do not yet know which one). Space within a galaxy is so vast, in comparison to the sizes of stars, that almost no direct impact between stars should have arisen from such a galactic collision. However, the gravitational force of the passing galaxy created "wakes" that drew together masses of gas, triggering bursts of star formation that produced the ringlike structure. The passage of the interloper through the galaxy, which is even larger than our own Milky Way, has caused something like a billion stars to be born. The blue "knots" in the ring are large clusters of new stars, some of which have been partially disrupted by supernovae (exploding massive stars). Photograph by Kirk Borne, Courtesy of NASA, Space Telescope Science Institute

Figure 7. In March 1994 the HST photographed this large elliptical galaxy in the Coma cluster, at a distance of about 300 million light years. The image shows some foreground stars in our own galaxy, and one or two other galaxies in Coma, but most of the visible galaxies lie far beyond the cluster. Courtesy of NASA, Space Telescope Science Institute

Figure 8 (overleaf). In this recent map of nearly 10,000 galaxies around the Milky Way by Margaret Geller and John Huchra, both the dark voids and the sheets of thousands of galaxies are larger than 100 million light years. The outer edges of these wedge-shaped slices lie approximately 400 million light years from the Milky Way. Distances to the galaxies have been assigned on the basis of their Doppler red shifts, rather than by direct measurement. A great goal of theoretical cosmologists is to reproduce these structures. Image by Margaret J. Geller, John P. Huchra, Luis A. N. da Costa, and Emilio E. Falco, Smithsonian Astrophysical Observatory © 1994

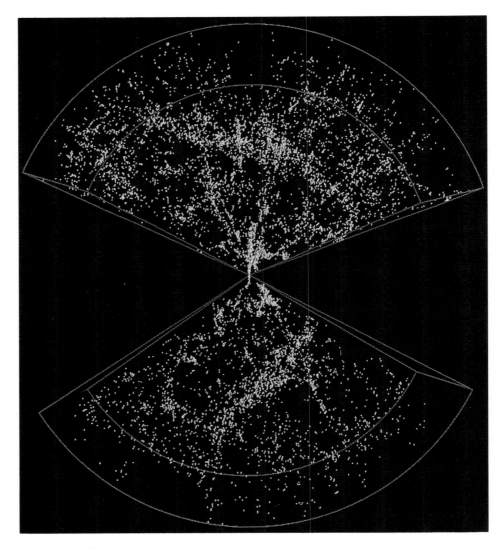

Figure 8. Distribution of galaxies. Legend on overleaf.

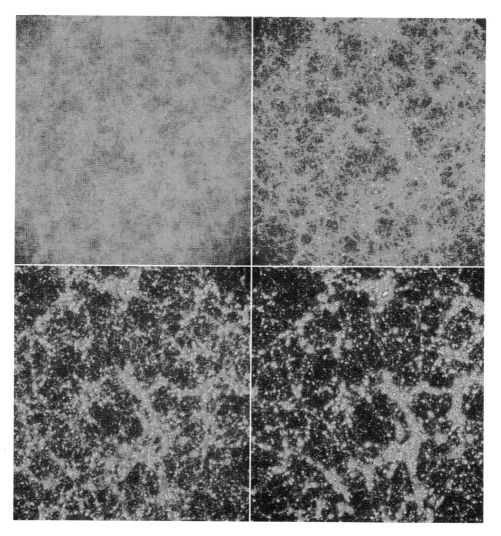

Figure 9. Computer simulation of galaxy formation. Legend follows.

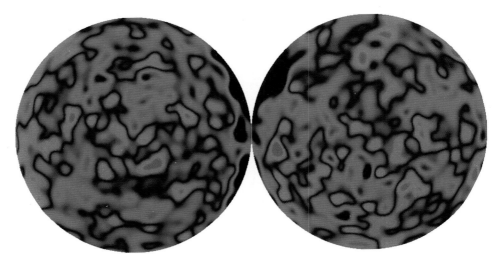

Figure 10. In 1992 cosmologists announced that they had found the first clear evidence of anisotropies—deviations from place to place—in the temperature of the cosmic background radiation. These deviations mark the sites of "seeds" for the large-scale structure that we observe today (see Figure 8). This recent COBE map shows the deviations in the two hemispheres of the sky, looking perpendicularly to the disk of the Milky Way. However, even on this map, much of the deviation (shown with different colors) consists of random noise; the actual anisotropies must be teased out of the data by careful statistical analysis. Courtesy of NASA/Goddard Space Flight Center

Figure 9 (overleaf). These panels show a computer simulation of the formation of galaxies. The simulation begins with an assumed distribution of deviations from absolute smoothness in the density, and proceeds from the top left (about one billion years after the big bang) through the top right and lower left to the lower right (the present time). (Blue represents the lowest density, which then rises through cyan, green, red, yellow, to white.) The panels expand along with the universe; the one at lower right has sides about 650 million light years long. Red, yellow, and white concentrations should be thought of as galaxies plus their much larger dark-matter halos. Comparison of these simulations with the actual distribution of galaxies in Figure 8 suggests that computers may yet model accurately the development of structure in the universe. Courtesy of Los Alamos National Laboratory, from work by Wojciech H. Zurek and Michael S. Warren

Figure 11 (opposite). The spiral galaxy Messier 100 belongs to the Virgo cluster, the closest large cluster of galaxies to the Milky Way. In 1994 photographs made with the Hubble Space Telescope revealed the brightest individual stars in Messier 100's spiral arms, some of which proved to be Cepheid variable stars. These Cepheids allowed astronomers to redetermine the distance to Messier 100 as approximately 52 million light years. The inset, which shows the central region of Messier 100, testifies to the impressive resolving power of the refurbished HST. Courtesy of NASA, Space Telescope Science Institute

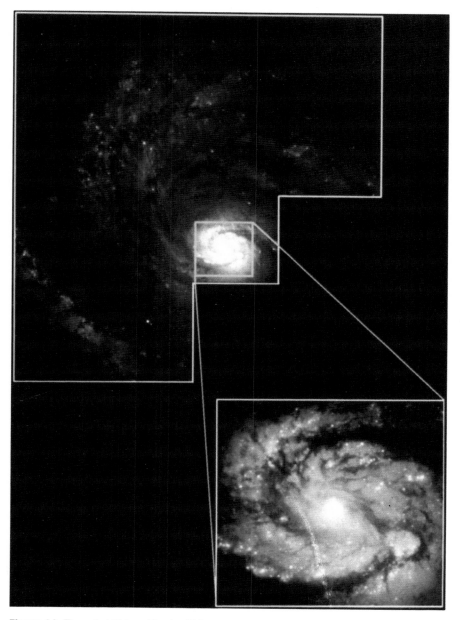

Figure 11. The spiral Galaxy Messier 100. Legend opposite.

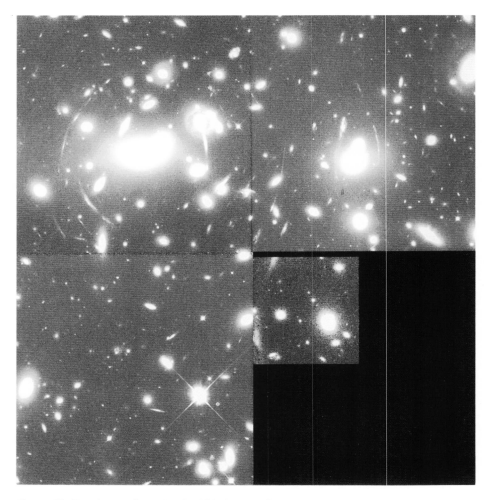

Figure 12. The cluster of galaxies Abell 2218, about 3 billion light years away, was photographed by the HST in September 1994. It shows the effects of gravitational lensing—the focusing of light by gravitational forces—in the form of gravitational arcs. These arcs arise from distortion of the light from galaxies far beyond the visible cluster; in some cases more than one image of a particular source of light is produced. Since greater gravitational forces create greater distortion, astronomers can learn much about the distribution and amounts of mass within a galaxy cluster by studying these gravitational arcs. Courtesy of NASA, Space Telescope Science Institute

The conclusion that the entire universe exists in a state of expansion leads to the notion that the universe has a definite origin in time, at least in its present state. If the universe is now expanding, then (by definition) its various parts used to be closer together. This implies that the density of the universe was higher in the past than it is now (see Diagram 5). If we run the cosmic movie backward, we reach a time when all the matter in the universe—and all the space too—was packed together at essentially infinite density—called the *initial singularity*. The big bang marks the moment when the expansion of space began.

In 1950 this moment was named the *big bang* in derision by the maverick astronomer Fred Hoyle, who was arguing in favor of an alternative model, *the steady-state universe*. According to this model, although the universe exists in a state of expansion, new matter continuously comes into existence at exactly the rate needed to maintain a constant average density of the universe. Thus, according to the steady-state model, the universe could have existed forever without changing its overall appearance.

In the third decade following Hubble's work, the 1950s, the steady-state theory vied successfully with the big-bang theory of the universe. No observations could definitively establish the validity of one model over the other. The most natural observations to distinguish between the two theories consisted of attempts to look far outward into the depth of space, and therefore far back in time, since light requires finite amounts of time to travel to us. By comparing the numbers of galaxies per billion cubic light years in regions relatively close to the Milky Way and regions billions

The Big Bang

of light years away, astronomers could hope to determine whether the average density of galaxies had changed with time.

The galaxy-counting method was hampered by the difficulty of making accurate observations of galaxies billions of light years away. More revealing information came from counting *radio galaxies,* a subclass of galaxies that emit far more energy in the form of radio waves than ordinary galaxies do. Galaxies emit primarily in the visible-light, ultraviolet, and infrared regions of the electromagnetic spectrum, because their stars produce their peak amounts of radiation in these parts of the spectrum. Stars do emit some radio waves, and so, therefore, do all galaxies. Some galaxies, however, release energy violently near their centers. This explosive release generates enormous amounts of radio waves, and makes the galaxy into what astronomers call a radio galaxy.

Using techniques developed during the 1950s, astronomers could detect radio galaxies at distances too vast for visible light to reveal much about them. They hoped that by counting radio galaxies they could succeed, where visible-light astronomers could not, in discriminating among competing models of the universe. The counts of radio galaxies gave strong indications that galaxies were more closely packed together in the past, and that therefore the big-bang model had a far greater claim to validity than the steady-state model. However, the steady-state model, like all theories, could be modified. Following their hearts (as scientists do), some cosmologists continued to support the steady-state model. This is the essence of science: Proponents of different theories argue as strongly as they can that "their" model surpasses all others in explaining what we see. In the case of the model universes, what was needed to resolve the dispute was observations of cosmic sources of radiation still farther away than the most distant radio galaxies.

Those observations soon became available. As we will see in the next chapter, the new evidence for the big bang consisted of electromagnetic radiation—streams of photons—that were created shortly after the big bang but were not detected on Earth until May 1964. They have been repeatedly observed since that time with improved equipment, most recently from a satellite dedicated to the task of studying these photons from the early universe. The Cosmic Background Explorer (COBE) satellite was launched in November 1989, and by January 1990 had made observations that confirmed beyond the skepticism of all but a few diehards the notion that the universe began with a big bang.

SIX

Looking for the Big Bang

In 1964, at the Bell Laboratories' Holmdel research station in New Jersey, a happy result for cosmology occurred. Arno Penzias and Robert Wilson, two astronomers employed by Bell Labs (an employment circumstance that brought astronomical credit to that giant company) were putting into operation a new antenna designed to gather and to focus short-wavelength radio waves, along with a new system designed to detect those waves.

Of all types of photons, radio waves have the longest wavelengths and lowest frequencies, but within the domain named "radio," some of the waves (today often called *microwaves*) have relatively short wavelengths and high frequencies. These short-wavelength radio waves are more affected by passage through the atmosphere than longer-wavelength waves and had therefore been little exploited for communication purposes. As a result, the design and manufacture of detectors capable of noting and analyzing the existence of such radio waves was in relative infancy, as were the antennas needed to gather them from a particular direction for study.

Penzias and Wilson found an annoying property of their new antenna and detector system: No matter how they tried to eliminate noise in the system, they seemed to detect a faint background, a modest amount of short-wavelength radio waves, in every direction of the sky toward which they pointed their antenna. They employed considerable ingenuity in eliminating possible sources of this noise, including careful cleaning of the pigeon droppings that had accumulated inside their antenna, but without success.

And then they learned that astronomers and physicists at nearby Princeton University had been trying, without success, to detect just this "noise"—which had already been predicted to exist as the faint whisper from the long-vanished early universe. Today astronomers call this whisper the cosmic background radiation.

The Origins of the Cosmic Background Radiation

The prediction that cosmic background radiation must exist was the brainchild of a Russian emigré physicist, George Gamow, and still more that of two of his students and associates, Ralph Alpher and Robert Herman. Gamow, born in Odessa in turbulent times (his father was a literature teacher whose dismissal was once urged by a group of students led by Lev Bronstein, later known as Leon Trotsky), managed to leave the Soviet Union legitimately in 1934, after an earlier attempt to cross the Black Sea in a collapsible kayak had failed miserably. He obtained a teaching position at George Washington University in Washington, D.C., where for a dozen years he considered problems that most physicists shunned as too intractable.

During the late 1940s Gamow grew interested in the question of how the universe formed the panoply of different elements we find around us today. If the universe has indeed been expanding since the moment of the big bang, when and where did it make its hydrogen, helium, carbon, nitrogen, oxygen? Did these nuclei emerge from the big bang itself, along with all the other types of nuclei, such as holmium, hafnium, and dysprosium? If so, could one calculate how much of each element would have formed through nuclear fusion processes during the first few minutes after the big bang, when the universe was incredibly hot and dense?

Gamow, Alpher, and Herman attacked this question. They assumed a universe initially made of neutrons, a first attempt at describing the early universe that yielded intriguing results. The physicists performed calculations to see what fractions of the different elements would emerge under the assumption of different initial conditions that described, among other things, the ratio of the numbers of photons to the numbers of neutrons and atomic nuclei.

From the calculations that Alpher, Gamow, and Herman made during the late 1940s, there emerged a peculiar conclusion, not much noticed at the time. If the early universe were to produce anything like the present relative abundances of the elements, it must have been amazingly rich in light and all other forms of electromagnetic radiation.

To appreciate this fact, we must recognize that the results of modern physics require us to view electromagnetic radiation not only as a series of *waves* but also as groups of *particles*. We must try to envision electromagnetic radiation as streams of massless bullets, each carrying energy and traveling at the speed of light, and also endowed with a wavelength and a frequency. From the Greek word for "light," physicists named these massless bullets *photons*.

Alpher, Gamow, and Herman calculated that for every neutron or atomic nucleus, the universe should contain at least a million photons. (Today we set that ratio closer to a billion.) These photons emerged during the first moments after the big bang, when the universe was enormously hot. Any matter hotter than absolute zero will continually emit photons; the higher the temperature, and the greater the amount of matter in existence, the larger will be the number of photons produced each second. Hence a hot universe will be a tremendous source of photons, which travel forever at the speed of light unless they collide with other types of particles.

Alpher, Gamow, and Herman saw that in terms of numbers of particles, the universe should consist mainly of photons, with everything else a small admixture. (Today we must add neutrinos and antineutrinos among the most numerous particles; see page 68.) Since photons have no mass, we "mass chauvinists" pay far more attention to particles *with* mass—neutrons, electrons, protons, and other types of nuclei such as helium, carbon, nitrogen, and oxygen. But the photons, dashing in all directions at the speed of light, dominate the universe in numbers, and have done so ever since the turbulent birth of the cosmos billions of years ago.

What types of photons are these? Could we hope to detect them? To Alpher, Gamow, and Herman, the answer to the first question was apparent. The cosmic sea of photons must now consist almost entirely of low-energy photons: radio waves. We have seen that photons can be characterized by a frequency and a wavelength, but they can equally well be specified by the amount of *energy* that they carry. The energy of any photon is directly proportional to the photon's frequency, that is, to the number of vibrations the photon makes each second. Hence the photon's energy must be inversely proportional to its wavelength, since for any photon, frequency times wavelength equals a constant, the speed of light.

The first few minutes after the big bang produced enormous numbers of high-energy photons, because matter in the universe then had enormous temperatures. Higher-temperature matter not only generates more photons per second than lower-temperature matter does; it also produces a larger ratio of high-energy to low-energy photons. During the first few minutes after the big bang, when the temperature ranged through the billions and millions of degrees, an (imaginary!) observer would have been immersed in a violent sea of high-energy gamma-ray photons, each carrying enough energy to be dangerous to all living creatures that we know.

With the passage of billions of years, the situation has changed. The high-energy gamma-ray photons that emerged in enormous numbers during the early minutes of the cosmos have become the low-energy radio photons that fill the universe today. How did this transformation occur?

The only good answer to the question, Where has the photon energy gone? is "Doppler theft," the robbery of energy by the expanding universe itself. Recall that all distant objects are receding from us, with the more distant objects receding more rapidly. (The fact that these objects may not now even exist is irrelevant; what counts is their motion at the time that they produced the photons that we now detect.) Quasars, for instance—the most distant individual objects yet discovered—have recession velocities that in some cases surpass 90 percent of the speed of light. Although quasars have sizes much smaller than galaxies (their name, an acronym for "quasi-stellar radio sources," testifies to their pointlike appearance), they rank as the most intrinsically intense sources of radiation in the universe. Many of them emit more energy per second in visible light, in infrared, or in radio waves than the total emission of a giant galaxy in all regions of the electromagnetic spectrum. Quasars may be the central regions of young galaxies, seen as they were many billion years ago; their tremendous energy output may arise from material falling into a supermassive black hole at the galaxy's core, heating and colliding violently on the way in.

To estimate the distances to quasars, astronomers rely on the Doppler shifts in their spectra. These spectra show familiar emission and absorption lines, but in patterns that are highly shifted toward the red (long-wavelength) end of the spectrum. The most straightforward interpretation of these Doppler shifts ascribes them to the expansion of the universe, which implies enormous velocities, and correspondingly enormous distances, to the quasars.

The Doppler effect has produced tremendous changes in the wavelengths and frequencies of the photons that we

Energy Loss through the Expansion of the Universe

detect from quasars. In many cases, photons that were emitted as ultraviolet photons reach us as infrared photons. The photons' frequencies have all been decreased, and their wavelengths all increased, by a factor of three or four. And since the frequencies that we observe equal only one-third or one-fourth the original photon frequencies, so too must the observed photon energies be only one-third or one-fourth of the energies with which the photons set out on their journeys. This is Doppler theft in action.

Another way to think of Doppler theft is this: the universal expansion continually increases all distances in the universe, and therefore continually stretches all the photon wavelengths. In this alternative view of the Doppler effect over cosmic distances, we can see that photons will take longer to travel greater distances, so the expansion of the universe will have more time to stretch the wavelengths for more distant sources. No one can recover the photons' lost energy; it has vanished forever, gone with the universal expansion like the wind that whirled across Georgia in 1864.

For the photons produced during the first few minutes after the big bang, a quasarlike recession velocity equal to a mere 90 percent of the speed of light would be excessively modest. For these photons, the relevant velocity equals approximately 99.9998 percent of the speed of light, just two parts in a million less than light speed itself. This recession velocity produces an enormous effect on the photon wavelengths, frequencies, and energies. The observed photon wavelengths are all a thousand times longer, and the frequencies and energies are all one one-thousandth as large, as they would be if no recession velocity existed.

And even this factor of one thousand understates the case, because it refers to what the Doppler effect has done during the time interval that begins not a few minutes but rather about a million years after the big bang. To understand why this interval in time plays a crucial role in the history of the cosmos, we must pause to consider how photons interact with the rest of the universe.

The First Few Minutes

The high-energy photons created by the big bang initially had a tremendous influence on the rest of the universe. If they encountered an atomic nucleus, they each had enough energy to break the nucleus into protons and neutrons. If they met an atom, they could knock the atom's electrons loose from the nucleus, giving each "free" electron an enormous amount of energy. But the expanding universe stole energy from every photon. If you were sitting on a nucleus in the early universe, you would find that after the first few minutes had passed, no photon would arrive with an energy sufficient to break the nucleus apart. Any photon arriving then would have covered many million kilometers to reach you. The photon from this "distant source"—once again the universe in its first few minutes—would have had its energy reduced by the Doppler effect to the point that it posed no further threat to the continuing existence of any types of nuclei.

Thus, after the first few minutes the basic mixture of nuclei in the universe had been established. As Alpher, Gamow, and Herman calculated, these were nearly all ordinary hydrogen (containing one proton); its rare isotopes,

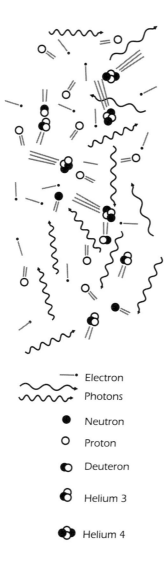

Electron

Photons

● Neutron

○ Proton

◐ Deuteron

❽ Helium 3

❽❷ Helium 4

deuterium (one proton and one neutron) and tritium (one proton and two neutrons); ordinary helium (helium 4, with two protons and two neutrons); and its variant, helium 3 (two protons and one neutron). All the other types of nuclei, all the carbon and oxygen, sulfur and silicon, iron, copper, nickel, lead, and zinc, all the gold, silver, mercury, and uranium, amounted to no more than one part in a million compared with hydrogen and helium. This tiny fraction results from the characteristics of helium nuclei, tough little devils that cannot easily be made to fuse into larger nuclei, because no stable nuclei exist with either five or eight nucleons (protons or neutrons).

All the elements with nuclei more complex than helium owe their creation to massive red giant stars, which fused the heavier nuclei before spewing them through space in supernova explosions. The history of element formation thus divides into two separate epochs. The first few minutes made the hydrogen and helium; everything else required "special" creation in fiery stellar furnaces and occurred randomly over an interval of time measured in billions of years.

If we focus on mass, the first few minutes after the big bang made significant amounts only of protons (hydrogen nuclei), neutrons, helium nuclei, and electrons. But if we focus on numbers of particles, we must add to this mixture enormous numbers of photons. Furthermore, we must include equally enormous numbers of neutrinos and anti-neutrinos, particles with zero or close-to-zero mass that emerged from early nuclear reactions but which are far less likely to interact with other particles than photons. We must also account for the fate of the neutrons that emerged in huge quantities from the first few minutes. Except for the neutrons that were locked into atomic nuclei, they have all decayed—fallen apart one by one to make protons, electrons, and antineutrinos. Neutrons by themselves cannot exist for long, but their survival is assured if they form part of a nucleus.

So a mass chauvinist surveying the universe half an hour after the big bang would find himself living in a cosmos in which protons, electrons, and helium nuclei are by far the most abundant identifiable particles. (In later chapters we shall discuss the "dark matter" with enormous amounts of mass that this chauvinist has missed.) But the imaginary observer could not ignore the massless photons. Even though no photon had enough energy left to break a nucleus apart, each photon retained sufficient energy to knock electrons loose from their orbits around nuclei. Any atom that happened to form as electrons swarmed around protons or helium nuclei therefore had a short life, broken by photon impact. As a result, the expanding universe was a froth of protons, helium nuclei, and electrons, interlaced with far greater numbers of photons, neutrinos, and antineutrinos—but without a single enduring atom.

The Era of Decoupling

Expansion itself allowed the universe to change from this near-featureless froth into the complex, structure-rich cosmos that we see today. This transition is called the *era of decoupling,* the time about a million years after the big bang when atoms found themselves able to form and to endure. Once again, the reason for the change lay in the expansion's

Before decoupling

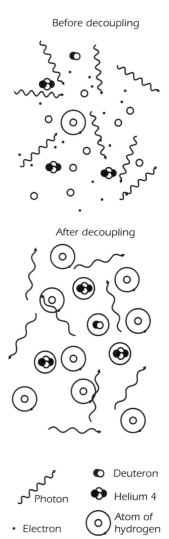

After decoupling

↜ Photon

• Electron

○ Proton

◐ Deuteron

◑ Helium 4

⊙ Atom of hydrogen

◑ Atom of helium

effect on photon energies. A photon needs far less energy to knock an orbiting electron loose from an atom than it does to fracture the nucleus of that atom. Hence photons retained their ability to destroy atoms for 300,000 years despite the lowering of all photon energies caused by the expansion. But when this time had passed, the photons that filled the universe found themselves just as unable to knock electrons loose from atoms as they had been to crack the nuclei themselves within a few minutes after the big bang.

As this inability to destroy atoms arose, the photons found themselves essentially unable to interact with hydrogen and helium atoms at all. The energy needed for any such interaction typically equals at least half of the energy needed to destroy an atom, so soon after the latter became no longer available, so too did the former. The photons could still interact with electrons that were floating freely in space, by passing some of their energy to the electrons through collisions. But before long, the abundance of these "free electrons" fell nearly to zero, as almost all the electrons in the universe began to orbit around either protons or helium nuclei, forming atoms of hydrogen and helium. Since these two types of atoms dominated the types of atoms in the universe (and still do!), the era of decoupling was the last time when the great numbers of photons in the universe had any cosmic interaction with matter.

To be sure, *some* interactions still occur between the sea of photons and particular types of atoms—or we would never be able to detect the photons. But no interactions occur between the photons and atoms of hydrogen or helium, which are by far the most abundant types of atoms in the universe. As a result, the photons produced throughout the universe soon after the big bang have passed throughout the universe with no significant change—except for Doppler theft—ever since the era of decoupling.

The factor of one thousand in energy loss that was quoted above refers to the effect of the expansion since the era of decoupling. Already by that time the Doppler effect arising from the expansion had changed the photons from gamma-ray to ultraviolet photons, as "seen" by any atom or nucleus about 300,000 years after the big bang. The total effect of the expansion since the time that the photons were originally created comes closer to one *billion*. Gamma-ray photons have become radio photons; if we seek to detect them now, we must use a radio telescope.

The Search for the Cosmic Background Radiation

Just after World War II, when Alpher, Gamow, and Herman made their original calculations, radio astronomy was in its infancy. The first detection of cosmic radio waves, from the sun and from regions of great cosmic violence, had just been achieved; radio emission from hydrogen atoms floating peacefully among the stars was first detected only in 1951. Alpher and Herman did discuss the possibilities for detecting the photons from the early universe with experts in radar from nearby universities and laboratories. They were told that the calculated number of photons, as well as the wavelengths that most of them should have, precluded the use of the radar systems then in existence for finding the cosmic sea of photons. In fact, this pronouncement was not correct, but it persuaded Alpher, Gamow, and Herman not to make further attempts to urge astronomers to look for the photons that their calculations predicted.

And so the calculations lay quietly in the journals. Though it may be difficult to accept today, no one devoted to observing the cosmos took theories of the early universe very seriously 45 years ago. The theories could provide useful employment for those immersed in such matters, but observational astronomers wanted real work. They had plenty to do, especially in the burgeoning field of radio astronomy, without worrying about whether calculations about the early universe provided reasonable grounds for their efforts.

Fifteen years after its publication, Penzias and Wilson at Bell Labs knew nothing of the work by Alpher, Gamow, and Herman. They mentioned their quirky result to Bernard Burke, a radio astronomer at MIT, who told them that "some guys at Princeton have predicted [a temperature] of 10 K in the X-band," the part of the radio spectrum under investigation. Penzias and Wilson promptly met with these guys: Robert Dicke, the inventor of the instrument that most astronomers used to measure the intensity of radio waves; James Peebles, a postdoctoral fellow who had calculated that if we assume that the early universe produced the helium that we find today, we should now find a cosmic sea of photons at a temperature of a few degrees above absolute zero; and David Wilkinson and Peter Roll, who were just then building a receiver system to search for this cosmic photon sea.

A few days' work convinced the Princeton scientists that Penzias and Wilson must be listening to the hum from the era of decoupling. Fifteen years earlier, Dicke's observations had set an upper limit on the strength of this hum; but since he did not find it, he had forgotten about this aspect of his research until Peebles found the published result in the library. The Princeton scientists had been listening for the cosmic hum with a small antenna mounted on the roof of the Princeton geology building. Finding no signal, they had realized that they could do better by observing with a larger antenna equipped with a more sensitive detector, something like the system a few miles away at Bell Labs. Their insight was acute—so much so that their idea had been (accidentally) carried out without their direct input.

In 1965 the news of the discovery at Bell Labs burst on the world of astronomy like a firebell in the night, as Thomas Jefferson might have said. For the first time, humans had found reasonably direct evidence that the big bang had occurred, in the form of the photons that were being continuously monitored by a radio antenna in New Jersey. Hubble had discovered that galaxies and galaxy clusters are receding *from us;* but astronomers required an additional hypothesis—that we have a representative (and not a special) view of the universe—to imply a big bang throughout the cosmos. The photons found by Penzias and Wilson were real and resembled those that would be expected from the time immediately following the big bang. Their spectral distribution corresponded to a temperature of 2.7 degrees absolute for the cosmic background radiation.

But were these photons really the relics of the first few minutes? Because most of the photons from the era of decou-

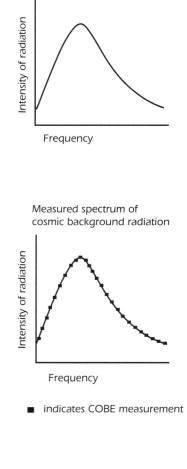

Predicted spectrum of cosmic background radiation

Intensity of radiation

Frequency

Measured spectrum of cosmic background radiation

Intensity of radiation

Frequency

■ indicates COBE measurement

pling cannot penetrate Earth's atmosphere, full confirmation of this conclusion was a long time coming—twenty-six years. In 1990 the COBE satellite measured nearly the entire relevant part of the spectrum of radiation, including the radio frequencies that cannot penetrate the atmosphere. The measured spectrum corresponds perfectly to what the big-bang model predicts. To most cosmologists, this confirmation was welcome but anticipated: They had already concluded that the photons first observed in New Jersey three decades ago are representative of the entire cosmic sea, produced during the first minutes of the expanding universe.

Because the type of radio waves that Penzias and Wilson observed are called microwaves, the cosmic sea of photons soon acquired the name *cosmic microwave background*—"cosmic" because the photons were believed to fill the universe, "microwave" because the observed photons have the frequencies characteristic of microwaves, "background" because the photons form a background to all other sources of photons in the universe. Today astronomers and cosmologists more generally use the name *cosmic background radiation* for the sea of photons that Alpher, Gamow, and Herman first predicted and that Penzias and Wilson first observed.

The discovery of the cosmic background radiation, and its later confirmation through improved observations showing that its spectrum exactly matches what a big-bang universe should contain, have left the big-bang model triumphant. And because the concept of a big bang has immense resonance with the human psyche, much has been made of its implications for theology (particularly as regards "creation") and eschatology (the branch of theology concerned with "last things," such as death, doomsday, heaven, hell). Does the big bang constitute scientific confirmation of the "creation" story described in Genesis and in the mythologies of many other cultures? Not necessarily. As we shall see in Chapter 10, our universe may be simply one of an infinite number of universes, in a much larger (and totally inaccessible) "metauniverse," in which individual universes pinch themselves off at random times and go their own way.

And what about those "last things"? Does the big bang model allow us to predict the future and fate of the universe as well as recreate its past? Quite possibly—if we can only learn a bit more about the contents of the cosmos.

SEVEN

Walls of Galaxies, Fingers of God

Without thinking about the immense difficulties involved, one might easily leap to the conclusion that astronomers must have determined the distances to at least many thousands of the millions of galaxies they can see. But a good estimate of the distance to an individual galaxy depends on finding at least one, and preferably several, Cepheid variable stars or other "standard candles." This task is not an easy one, and the number of galaxies with well-determined distances remains surprisingly small.

As we saw in Chapter 3, Cepheids can be seen and recognized only out to the Virgo cluster, the large galaxy cluster closest to our own Milky Way. If we hope to estimate the distances to the myriad other galaxies that surround our closest neighbors, we must find standard candles that are intrinsically more luminous than the Cepheid variables. Such objects exist, though they cannot be calibrated as accurately as the Cepheids.

Supernovae—exploding stars that flash out unexpectedly, then briefly shine with the luminosity of a billion suns before slowly fading into obscurity—can furnish what we seek, so long as we can deal with the fact that they do not all behave identically. The following chapter describes these standard candles in more detail; for now, we can simply analogize the use of supernovae as distance indicators to the Cepheid-variable approach. Once again, astronomers make careful measurements of the apparent brightnesses of a class of objects thought to be intrinsically much the same, but located at different distances from us, in order to estimate the ratios of those distances. The use of supernovae as distance indicators has only recently become feasible, thanks to an increased understanding of just how stars explode. This method promises, however, to furnish just what astronomers have long sought: standard candles visible out to distances of hundreds of millions of light years.

Of course, speed in scientific research is not always possible. The lesson learned from attempts to map the cosmos is a familiar but important one: To make a scientific breakthrough, scientists must be prepared to spend many years of hard work without much recognition. Barbara McClintock worked on genetics for years before her insights into "jumping genes" were acknowledged; and one of her predecessors, Gregor Mendel, spent his entire career utterly unknown (not so surprising, since he published only in obscure journals and wrote in Czech) before his discovery of the gene phenomenon was posthumously appreciated. Long hours of labor in obscurity will themselves not assure any outstanding result, but they are a prerequisite to great accomplishment in all but the rarest of cases.

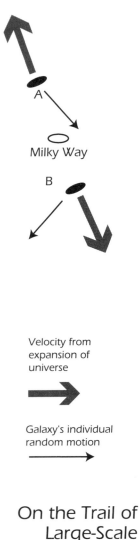

A

Milky Way

B

Velocity from
expansion of
universe

Galaxy's individual
random motion

The astronomers most directly involved in long-term projects to map the universe are two Harvard professors, Margaret Geller and John Huchra. Geller and Huchra were already interested in observing faraway galaxies when they met during the late 1970s. Both of them were then postdoctoral fellows at the Smithsonian Observatory, which has long been affiliated with Harvard College Observatory in a symbiotic relationship. The Smithsonian owns and operates several telescopes in Arizona, and one of these was available for long-term surveys, as part of a project that had been set in motion by another Harvard astronomer, Marc Davis. Geller and Huchra decided to use the telescope to map the galaxies surrounding us in what we might call the nearby cosmos—farther than Davis had looked, but still far short of the greatest galaxy distances.

Since Geller and Huchra wanted to map the positions of many thousand galaxies with respect to our own, they required a method that would yield this result within a few years. That method lay conveniently to hand: Measure the Doppler shifts in the spectrum of each galaxy, and compute its distance using Hubble's law, $V = H \times D$. Actually, since the value of Hubble's constant H is uncertain (see Chapter 8), Geller and Huchra contented themselves with measuring redshifts to determine V, and left the distance estimates to astronomers more certain of the value of H.

On the Trail of Large-Scale Structure

Hubble's law states that any galaxy's recession velocity is proportional to its distance from us. Since $V = H \times D$, if we can determine the value of H, the Hubble constant, then we can find any galaxy's distance from us by *dividing* its recession velocity (measured from its Doppler shift) by Hubble's constant: $D = V/H$.

Unfortunately, determining the value of Hubble's constant is far from easy. Furthermore, this method rests on the assumption that every galaxy's speed of recession obeys Hubble's law, but we know that this is not quite true: Every galaxy has some individual random motion that will add to, or subtract from, the recession velocity arising from the expansion of the universe. If we measure a galaxy's recession velocity and insist that all of it arises from the universal expansion summarized by Hubble's law, we inevitably introduce some error in deriving the galaxy's distance.

The good news is that since the amount of galaxies' random motions equals "only" a few hundred kilometers per second, when we examine progressively more distant galaxies, with recession velocities of first a few thousand and then many thousand kilometers per second, the random velocity component introduces a progressively smaller difference between any galaxy's actual distance and the distance that we estimate by assuming its random velocity to be zero (see drawing at right).

There is another piece of good news embodied in the method that Geller and Huchra used to make their survey of the cosmos. Because they had full-time use of the telescope, they could arrange to observe the spectra not of one or two galaxies in a night but of a dozen or more. Of course, each spectrum still must be carefully measured to find the familiar pattern of absorption lines and to determine the amount by which the wavelengths and frequencies of those lines differ from their laboratory values. But the survey offered the promise of a dozen or more galaxy distances per

clear night—thousands in a year, tens of thousands in a decade.

Since Geller and Huchra knew that even several years' work could cover only a small fraction of the sky, they chose to survey a long, narrow strip, 6 degrees wide and 120 degrees long, running one-third of the way around the celestial sphere. Their decision was based on the fact that choosing a long, narrow strip would reveal more about the characteristics of galaxies' distribution in space than would a survey of one particular area of the sky that covered the same total area. For example, a long, narrow strip almost anywhere on Earth's surface will reveal both continents and oceans, whereas a survey of a square region will typically show either continents or oceans, but not both.

The method that Geller and Huchra employed to survey a strip around the sky allowed the Earth's rotation to carry the galaxies chosen for study into the telescope's field of view, with small motions of the telescope to adjust for successive galaxies entering the field. The light from each galaxy passes through a spectrometer, where after 5 to 10 minutes it produces a revealing smudge: the spectrum of the galaxy, ready to be measured to reveal the amount of the Doppler shift, and thus the galaxy's recession velocity. A simple division of this number by Hubble's constant then yields the estimate of each galaxy's distance—an estimate that will change if the value astronomers determine for Hubble's constant changes.

Once Geller and Huchra had completed a map of one slice of the sky, they began on the adjacent slice. Map a sufficiently large number of slices, and you will map the entire sky visible from a particular observatory. By now, Geller, Huchra, and their collaborators have mapped a dozen slices of the sky from both the northern and southern hemispheres of Earth (see Figure 9). They have measured the redshifts of more than 10,000 galaxies.

What the Slices of the Sky Revealed

Even the first slice produced in 1986 revealed a map strikingly different from what astronomers had anticipated. Later results have amply confirmed those first, shocking views: The distribution of galaxies is stunningly nonrandom, nonuniform, and highly complex.

Astronomers had long known that galaxies tend to cluster together in groups that include several hundred to a few thousand members. These clusters, of which the Virgo cluster is the closest example, typically span distances of 10 or 20 million light years. But no one anticipated that on much larger distance scales—50 to 1,000 million light years—the universe would reveal a structure far more complex than a galaxy cluster, a delicate spongelike web of strands and sheets that surround enormous regions nearly devoid of galaxies. Most impressive of all in the Geller-Huchra maps is the Great Wall of Galaxies that extends almost all the way across the map, spanning close to a billion light years.

Geller and Huchra produced their first maps, showing just a few slices of the cosmos, nearly a decade ago. Since then, they have continued to map more and more of the cosmos surrounding the Milky Way (Figure 8). Because they observe only those galaxies with the greatest apparent bright-

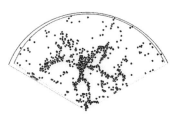

nesses, their survey extends outward for "only" about 400 million light years. Because structures such as the Great Wall of Galaxies are nearly as large as the map itself, we have yet to answer the question, How large are the largest structures in the universe? Until now, each time that the cosmos has been mapped at progressively larger distance scales, structures as large as each new map have been found.

The larger the structures that astronomers find, the greater will be the difficulty in explaining them. The reason for this is the relatively short amount of time that has elapsed since the big bang. As we shall see in Chapter 13, astronomers encounter difficulties explaining how even individual *galaxies* formed in the allotted time. The problem of creating any kind of organized structure in the universe within a specified period of time grows more acute as we look to progressively larger structures, because the formation of a galaxy requires moving material over much shorter distances than must be involved in the formation of far larger structures. If astronomers have trouble explaining the existence of galaxies, they have still more difficulty in explaining the Great Wall of Galaxies.

Pointing Fingers

Geller and Huchra's map has been reproduced hundreds of times, and Margaret Geller and the filmmaker Boyd Estus have produced a marvelous movie ("So Many Galaxies . . . So Little Time") that uses computer graphics to sail through the cosmos revealed by the survey. However, if you want to impress your friends and relatives with your knowledge of this map, perhaps the surest way is to point out that many of the most "obvious" features in this map are in fact illusions! These are the "fingers of God," named by astronomers as a jest, since astronomers understand the difference between human-made and divine artifacts. The "fingers of God" are the apparent lines of galaxies that line up with the point of origin—our own Milky Way, located at the vertex of the map of the slices that extend in opposite directions (the cosmos as mapped from the northern and southern hemispheres of Earth).

What are these "fingers of God" that point, providentially, to our own galaxy? Fittingly enough, they arise from the all-too-human assumption that *all* of a galaxy's recession velocity arises from the expansion of the universe. Astronomers know quite well that this is not so, but since we do not know the individual, random component of any particular galaxy's motion, a reasonable first attempt to map the universe uses the assumption that each galaxy has a random velocity component of zero.

To understand how this artifact arises in the map, look first at the striking concentration at the center of the upper half. This is a cluster of galaxies called the Coma cluster because its center lies in the constellation Coma Berenices (Figure 7). The few thousand galaxies that belong to this cluster (like all galaxies in all clusters) are held together by their mutual gravitational attraction, which causes them to orbit around the cluster's center. Their velocities in orbit do *not* arise from the universal expansion, but rather fall in the other category of velocities, that of galaxies' random velocities. At any moment, some of the galaxies in the Coma cluster have velocities of approach, and a roughly equal number have velocities of recession, in comparison with the velocity of the center of the cluster. These galaxies are in fact not moving exactly toward us or exactly away from us, but their motions include components of approach or recession, and it is this component that we measure with the Doppler shift in the spectrum of the light from each galaxy.

Geller and Huchra make maps that plot simply the galaxies' Doppler shifts, not their distances. If we assign distances to galaxies solely on the basis of their measured recession velocities, then we shall assign overlarge distances to galaxies whose motions within the cluster happen to be carrying the galaxies away from us, and we shall underestimate the distances to galaxies whose orbital velocities include a component of approach. This spreads the Coma cluster in the map outward from what would be a dense, roughly spherical dot into the elongation that we see, one of the "fingers of God," which all share the characteristic that their elongated forms point directly at the origin of the map.

The "fingers of God" provide an excellent reminder that when astronomers (or other scientists) present their data, they do so in ways designed to be highly useful to *them,* but which include important assumptions not immediately apparent from their presentation. Once you know about the "fingers of God," you can easily adjust your interpretation of the Geller-Huchra map, whose basic impact remains unchanged: The cosmos has a complex structure that defies any explanation.

Cosmic Flows and the Great Attractor

Some astronomers have attempted to use their analyses of galaxies' redshifts to search for clustering in the distribution of velocities that could reveal enormous regions with a higher density than average. These regions would reveal themselves through the gravitational forces that they exert on the galaxies in the maps. Additional large amounts of gravitational force would create an additional "flow" in velocity, superimposed on the basic "Hubble flow"—the expansion of the universe.

During the past decade, astronomers have begun to distinguish these large-scale flows from the cosmic expansion. To do so demands an effort barely achievable in modern cosmological research, since it requires complex statistical analysis of large numbers of galaxies, with a measured Doppler shift and an estimated distance for each one. In addition, the analysis must deal with the fact that each galaxy has some individual motion (see page 79), which simply confuses the attempt to determine whether significant numbers of galaxies in a particular large region possess a general flow that deviates from the Hubble flow.

But the effort offers an immense reward: Any large-scale flows, embracing galaxies that span a few hundred million light years in distance, can tell astronomers—at least roughly—how much mass produces those flows. The greater the concentration of mass, the greater will be the flow velocity determined by analyzing the motions of hundreds or thousands of individual galaxies.

Furthermore, the flow velocity responds to all types of mass, including the dark matter whose unknown amount determines the future of the universe. If we can distinguish and measure a flow, our determination of the density of matter will apply not to volumes as small as a galaxy cluster (of the order of 10 million light years across) but rather to regions hundreds of millions of light years in diameter. The flows can reveal the total density of matter on the largest distance scales that we can hope to measure (except for the hope of determining the average density of matter in the entire universe). Cosmologists would therefore delight in learning how galaxies flow on large distance scales. The present flow results, though mixed and controversial, lead to an important cosmological conclusion.

At this stage in our understanding of the universe, several regions of the visible universe have been reported to have large-scale flows, though none has been unequivocally accepted. By far the most famous of these flows arises from the Great Attractor, the increase in density claimed to produce deviations from the Hubble flow among large numbers of galaxies, each a few hundred million light years from the Milky Way, that are visible in southern-hemisphere constellations such as Scorpio and Centaurus. Proud of their accomplishment, the discoverers of the Great Attractor have given themselves an equally winning name, the Seven Samurai. Like the samurai of old, these astronomers have not failed to encounter opponents who dispute their interpretation of the data. At this time, an outsider would do well to stay tuned for further news of the Great Attractor and its champions, rather than accepting the Great Attractor on the same footing as the Great Wall of Galaxies.

But a wider-ranging analysis of galaxies' motions, which does not attempt to locate a particular concentration of mass such as the Great Attractor, now seems to show clearly that galaxies' motions do deviate from the Hubble flow far beyond the results of their individual random motions. Even if we cannot specify the details of any particular flow (because the Doppler effect reveals only velocities along our line of sight, and cannot show just which galaxies are headed in particular directions), we can conclude that flows do exist, and that to produce them requires enormous amounts of matter. In Chapter 11, we shall discuss the question of just how much matter the flows reveal; for now, we may enjoy the fact that cosmic flows on large distance scales offer the possibility of determining the density of matter in the universe, whether or not that matter emits light or any other type of electromagnetic radiation.

Even Larger-Scale Structure

In recent years, the type of survey that Geller and Huchra pioneered has been extended to still greater distances by a team of astronomers led by Robert Kirshner of Harvard, Augustus Oemler of Yale, Paul Schechter of MIT, and Stephen Shectman of the Carnegie Observatories. In order to measure the redshifts of a large number of galaxies on each night of observation, the astronomers carefully drilled holes into metal plates at positions that corresponded to the galaxies' locations on the sky, and attached fiberoptic cables to the holes. This technique allows them to measure the Doppler shifts for a hundred galaxies simultaneously.

By now, these astronomers, aided by Douglas Tucker and Huan Lin—graduate students at Yale and Harvard, respectively, who did much of the most difficult work—have measured redshifts for 25,000 galaxies in six strips along the sky, each 1 1/2 degrees wide and 75 degrees long. Their survey reaches out to distances four times greater than the limit of the Geller-Huchra slices. The farther-reaching survey reveals large voids and sheets of galaxies, on size scales similar to those seen in the earlier surveys. However, the survey does not clearly show any structures such as bubbles and sheets much larger than the ones already identified. Although there are intriguing hints that larger structures may reveal themselves upon more detailed analysis, it may turn out that astronomers have finally surveyed the cosmos to sufficient depth that they have found its largest structural patterns.

Even with this happy result, astronomers have their hands full in explaining how structures as large as a billion light years across could have formed during the time since the big bang (see Chapter 13). This problem has grown slightly worse with the decrease in the estimated age of the universe. It is time to face the problem head-on, and to examine the debate over the new values for the interval since the cosmos began its expansion.

Diagram 1. Visible matter. Legend follows.

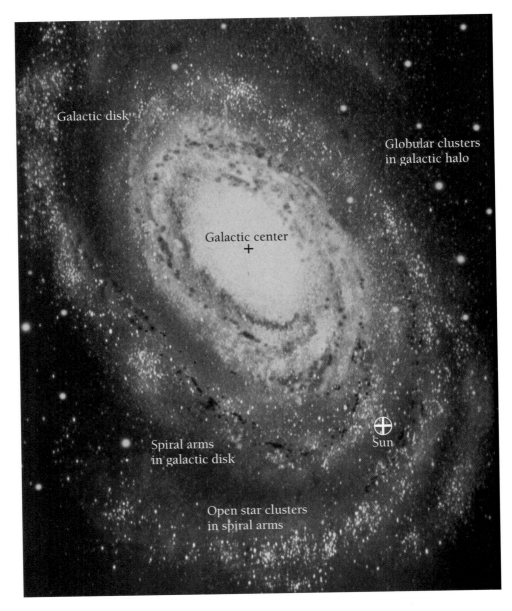

Galactic disk

Globular clusters
in galactic halo

Galactic center
+

Spiral arms
in galactic disk

Sun

Open star clusters
in spiral arms

Diagram 2. The Milky Way Galaxy. Legend follows.

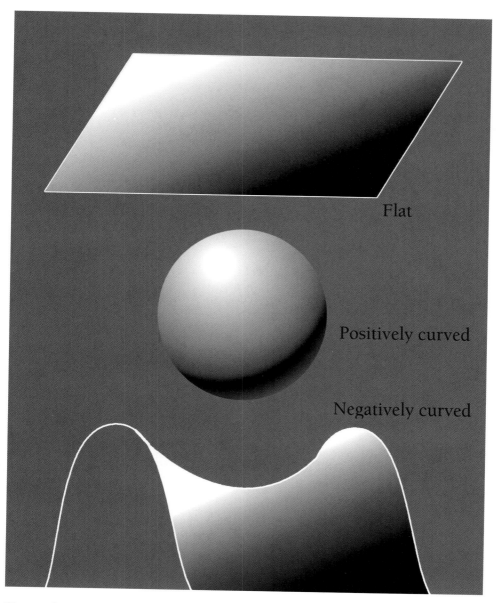

Flat

Positively curved

Negatively curved

Diagram 3. Three shapes of space. Legend follows.

Diagram 1. The visible universe consists of stars in galaxies, which themselves group together to form galaxy clusters. The distribution of stars gives most galaxies either spheroidal or ellipsoidal shapes (the elliptical galaxies) or else a highly flattened disk (the spiral galaxies), in which the youngest and brightest stars outline the galaxies' spiral arms.

Diagram 2. The Milky Way is a typical giant spiral galaxy, in which the stars form a flat distribution approximately 100,000 light years across but only a few thousand light years thick. Our solar system lies about 30,000 light years from the galactic center, near but not inside one of the Milky Way's spiral arms. Surrounding the disk is an extended halo of stars and globular clusters, apparently left behind as the galaxy contracted to its present size.

Diagram 3. Space in the universe can be described as either flat, positively curved, or negatively curved. Flat space corresponds to our intuitive feelings: It extends uniformly to infinity in all directions. In contrast, positively curved space bends back upon itself, like the *surface* of a sphere. (The diagrams shown here represent two-dimensional analogs of the possibilities for three-dimensional space.) Negatively curved space curves one way in a particular direction and the opposite way in the perpendicular direction; like flat space, it extends to infinity without curving back upon itself. Only one of these three possibilities accurately describes the real universe; unfortunately, we do not know which one that is.

Diagram 4 (opposite). Astronomers often plot the luminosities of stars (their energy output per second) versus the stars' surface temperatures to obtain a temperature-luminosity diagram. For historical reasons, surface temperature increases to the left in the diagram. Most stars belong to the main sequence, which sweeps from top left to bottom right in the diagram; lesser numbers of stars are either young hot blue giants (top center and left center), red giants that have left the main sequence (top right and right center), or still more highly evolved white dwarfs (lower center). The diagram's axes also show the spectral classes assigned to stars on the basis of their surface temperatures, as well as the stars' absolute magnitudes—their luminosities according to an ancient system in which smaller numbers indicate greater intrinsic brightness. The sample of stars shown here overrepresents the more luminous stars; a true sampling would reveal enormous numbers of white dwarfs and low-luminosity main-sequence stars.

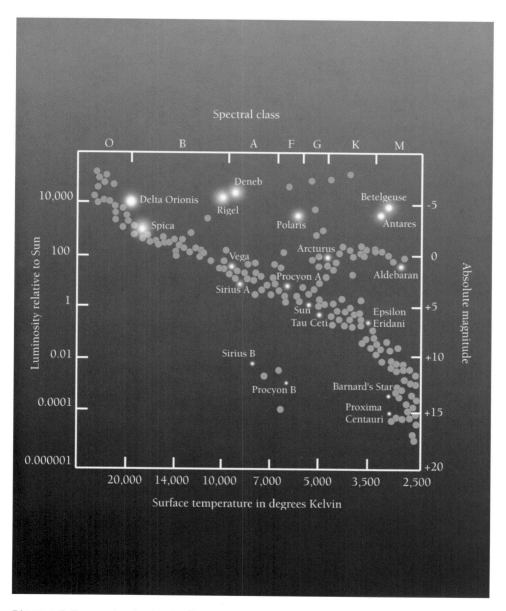

Diagram 4. Temperature-luminosity diagram. Legend opposite.

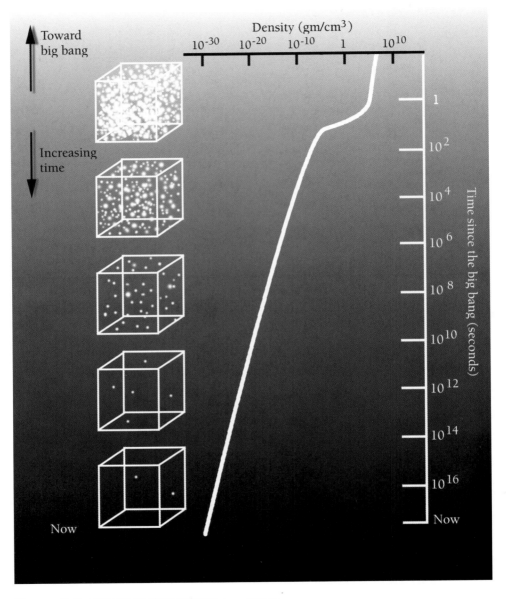

Diagram 5. Density history of the universe. Legend follows.

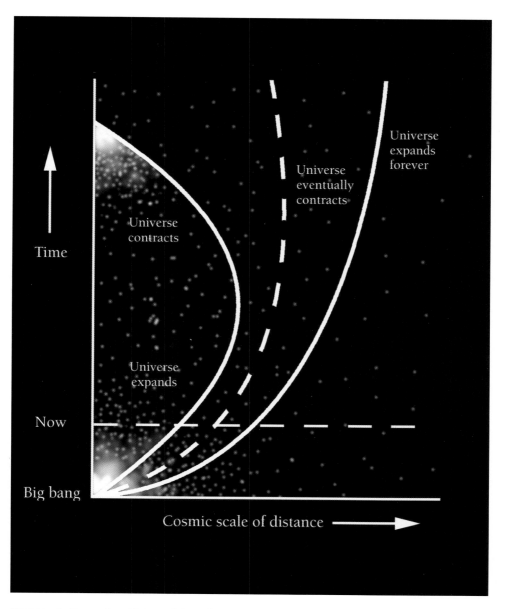

Diagram 6. The past and future of the universe. Legend follows.

Diagram 5. Because the amount of mass in the universe has not changed significantly since the big bang, the average density of matter (mass per unit volume) has steadily decreased as the universe has expanded. The graph shows the changes in the average density during the time a fraction of a second after the big bang. Because the time scale is logarithmic, the big bang itself lies an infinite distance above the top of the page. The kink in the curve showing the average density arose when the universe changed from having most of its energy in the form of radiation to having most of its energy in the form of energy of mass ($E = mc^2$).

Diagram 6. To show the possibilities for the future of the universe, we can plot the average separation between two representative points (the cosmic scale of distance) as a function of increasing time. The universe, which is expanding now, will either (1) expand forever, (2) cease its expansion after an infinite amount of time has passed, or (3) eventually contract. The first possibility corresponds to a negatively curved universe, the second to a flat universe, and the third to a positively curved universe (see Diagram 3). As a practical matter, the first two possibilities are nearly identical.

Diagram 7 (opposite). When astronomers observe the cosmic background radiation from any two regions of space located in opposite directions, they find that it has almost exactly the same characteristic temperature. This uniformity implies that the entire universe had nearly identical characteristics a few hundred thousand years after the big bang—the time when the background radiation we observe today last interacted with matter. But even at that time, regions of space in opposite directions from what became the Milky Way were already too far apart for information to flow between them, since no information can travel more rapidly than the speed of light. Cosmologists call this uniformity of the cosmic background radiation, which is difficult to explain, the "horizon problem."

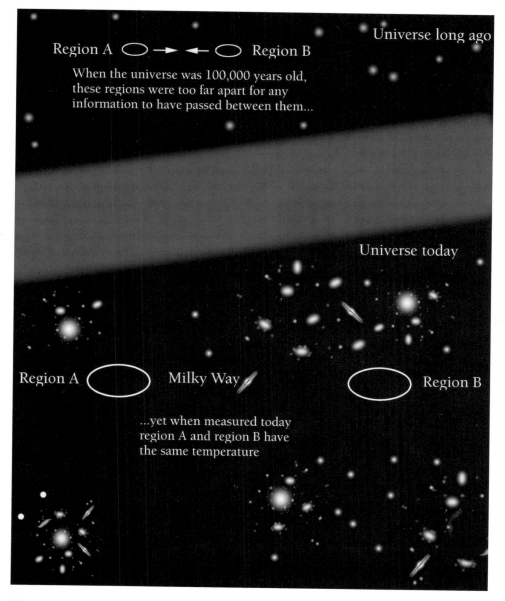

Region A ⬭ → ← ⬭ Region B

Universe long ago

When the universe was 100,000 years old, these regions were too far apart for any information to have passed between them...

Universe today

Region A ⬭   Milky Way ⬭   ⬭ Region B

...yet when measured today region A and region B have the same temperature

Diagram 7. The horizon problem. Legend opposite.

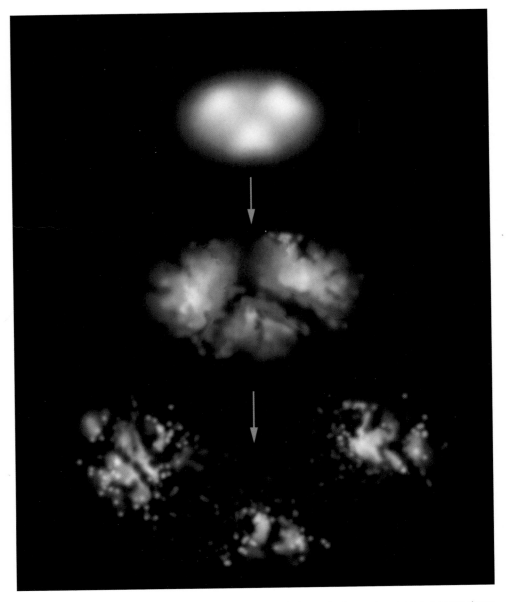

Diagram 8. In "top-down" models of how matter formed clumps in the early universe, a large clump of matter fragments into smaller clumps, which in turn fragment to form still smaller ones.

Diagram 9. In "bottom-up" models of how matter formed clumps in the early universe, small clumps of matter agglomerate into larger units, which in turn form still larger ones.

Diagram 10 (opposite). By comparing the measured abundances of different types of nuclei with the amounts calculated in models of the early universe, cosmologists can determine the present density of baryonic matter. They can do this because the density of baryonic matter today is proportional to its density during the first few minutes after the big bang, when nuclear fusion occurred throughout the universe. Because only baryonic matter participates in nuclear fusion, this method can yield only the density of baryonic matter. In this diagram, the gray bars show the measured amounts of different types of nuclei, and the black lines show the amounts calculated for different densities of baryonic matter. The best fit between the observations and the models occurs if the present density of baryonic matter is approximately 6×10^{-31} gram per cubic centimeter, though this number could be in error by nearly a factor of 2. The best-fit number is only about 4 to 8 percent of the critical density of matter that is needed to produce a flat universe, as required by the inflationary theory. However, the density of baryonic matter substantially exceeds the density of visible matter (see Diagram 11). This implies that the universe consists mostly of dark matter, which must be either mainly baryonic (if the actual density is less than one-tenth of the critical density) or mainly nonbaryonic (if the actual density is much greater than one-tenth of the critical density).

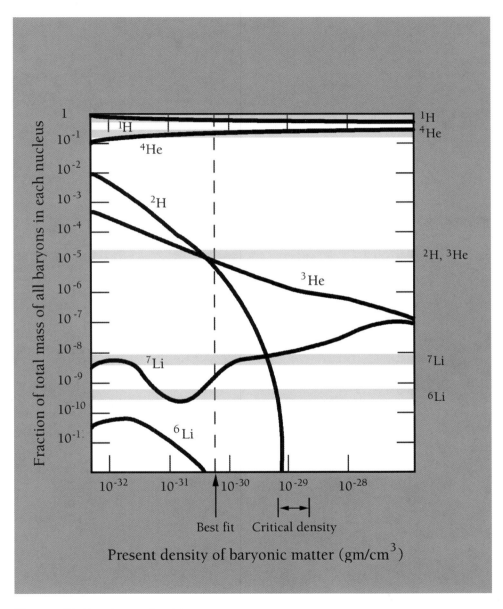

Diagram 10. Abundances of nuclei. Legend opposite.

Diagram 11 (opposite). This diagram shows current limits on the values of the Hubble constant, H_0, and on the total density of matter in the universe. The vertical lines at $H_0 = 40$ and $H_0 = 80$ are the limits on H_0 that many astronomers accept as valid. The broad dark lines show the density of visible matter (the absolute lowest possible density), the density of baryonic matter inferred from measurements of the abundances of different types of nuclei (see Diagram 10), and the limit imposed by assuming that the universe must be at least 10 billion years old and that the cosmological constant is zero. An additional limit (not shown) arises from observations of galaxy clusters, which strongly imply that the density of matter is at least 10 percent of the critical density. If we believe that all of these limits are valid, current observations restrict the value of the Hubble constant and the ratio of the density of matter to the critical density to roughly the light-gray trapezoidal area at the upper center. If H_0 proves to be substantially greater than 40, this region will shrink markedly, and if H_0 turns out to be more than 60, either the density of matter can no longer be equal to the critical density, or the cosmological constant much be something other than zero. After a diagram by David Schramm.

Diagram 12. Spiral galaxies such as our Milky Way each possess an enormous halo of dark matter, which extends far beyond the halo of old stars surrounding the galaxy. A typical dark-matter halo may have a radius 10 to 20 times the radius of the disk, and thus reach half way to the next galaxy. The halo drawn here would in fact be invisible at all frequencies and wavelengths of visible light.

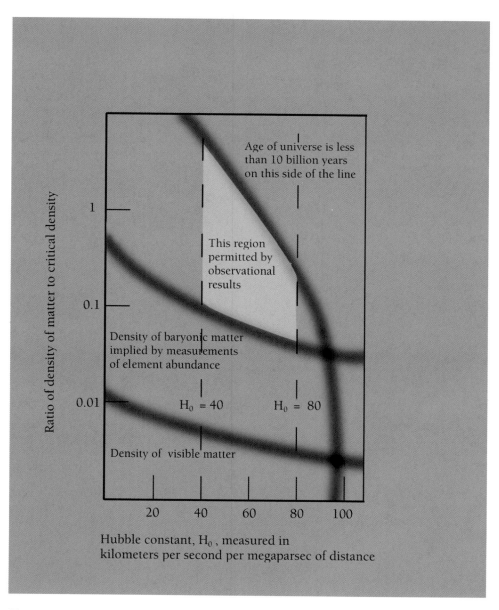

Ratio of density of matter to critical density

Age of universe is less than 10 billion years on this side of the line

This region permitted by observational results

1

0.1

Density of baryonic matter implied by measurements of element abundance

0.01

$H_0 = 40$

$H_0 = 80$

Density of visible matter

20 40 60 80 100

Hubble constant, H_0, measured in kilometers per second per megaparsec of distance

Diagram 11. Current observational limits. Legend opposite.

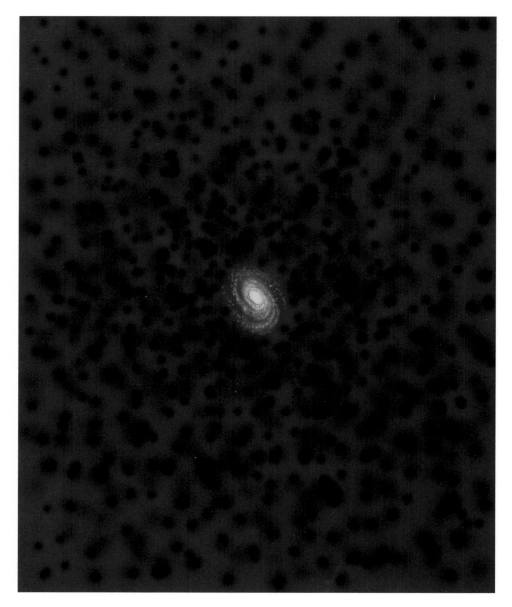

Diagram 12. The dark matter halo of a spiral galaxy. Legend precedes.

EIGHT

The Elusive Age of the Cosmos

Hubble's estimates of the distances to galaxies, combined with other astronomers' measurements of the galaxies' recession velocities, led astronomers to conclude that the entire universe—all of space and everything in it—is expanding, and that the expansion began at a definite moment in time, which we call the big bang.

If the big bang really did occur, then most people feel entitled to ask: Where did it happen? When astronomers lecture about the big bang, they have a bad tendency to cup their hands, as if the big bang happened "over there." This error highlights our natural tendency to try to imagine the universe as an object outside ourselves. In fact, the big bang occurred *everywhere,* at a time before the universe had any structure at all. So if you plan to act out the movie of time in reverse, be sure to put one hand above your head and the other below your socks before you squeeze them together.

Modern quantum physics implies that we cannot hope to know anything about conditions in the universe at times earlier than the "Planck time," about 10^{-43} second after the big bang. For most purposes (aside from the question of how the universe was created, the answer to which might well be found during the Planck time), it proves quite sufficient to imagine that the big bang began at that time, when the universe had already begun expanding.

In addition to the Planck-time difficulty, another barrier to easy understanding of the big bang concerns its size. The universe may be finite, or it may be infinite. In the former case, the big bang occurred when the universe occupied only a tiny volume; but if the universe has an infinite size today, then it was always infinite, and the big bang must have taken place within an infinite volume, albeit one that has expanded ever since the big bang.

These mind-boggling problems are details. The primary fact about the big bang remains that the big bang occurred at some finite time in the past. Thus, amidst their ethereal talk of finite versus infinite universes, Planck times, and the unknowable moment of creation, astronomers can (in theory) tell us the moment in time when the universe began its present state of expansion. The date of the big bang furnishes us with one hard fact to grasp within the slippery morass of shifting cosmological hypotheses.

Hubble's Law, the Key to the Age of the Universe

How do we know the age of the universe, the time since the big bang? In Chapter 5, we noted the relationship between galaxies' distances and their recession velocities. This relationship, called Hubble's law, describes the universal expansion; it states that galaxies are receding at speeds (which we denote by V) that are proportional to their distances from us (denoted by D), according to the simple equation $V = H \times D$, where H is a constant throughout the universe, called Hubble's constant. By calling H a "constant throughout the universe," we adopt the assumption that any observer, located anywhere in the universe, will now see what we see: galaxies receding from that observer with the same constant of proportionality between the galaxies' recession velocities and their distances that we observe.

What are the units of H? Is Hubble's constant a velocity, a distance, or their ratio? To answer this question more clearly, we may rewrite Hubble's law by dividing both sides of the equation by H. Then we obtain $D = V/H$, which shows that galaxies' distances are proportional to their recession velocities times a constant, $1/H$. Since distance equals velocity times time, $1/H$ must be measured in the units of time, and H must have units of "inverse time."

In fact (sparing the reader a discussion that might seem technical), the constant $1/H$ turns out to be exactly the amount of time that has elapsed since the big bang, *if* the rate of expansion has not changed. Hence, if we can determine the value of H, by measuring galaxies' distances and their recession velocities accurately, then H divided into one provides the time since the big bang, so long as we feel confident that the universe has not changed its rate of expansion.

However, the rate of expansion has certainly changed, and by an unknown amount. Gravity, which attracts all matter in the universe to all other matter, tends to slow the expansion. Astronomers therefore use the notation H_0 for the value of the Hubble constant now, aware that it was different

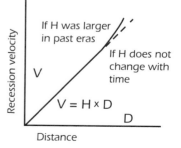

in the past. Since $V = H \times D$, if the expansion proceeded more rapidly in past eras, H must have been larger then. A larger H would match a given amount of distance between galaxies with larger velocities—more rapid expansion—than is now the case.

Attempts to date the big bang therefore have at least two fundamental problems. The first and probably the greater problem lies with estimating the distances to galaxies, and thus with obtaining a value for H_0, the Hubble constant that characterizes the universe now. The second problem resides with using H_0 to estimate the time since the big bang, that is, with attempting to figure out how H has changed.

The second problem can be well limited. The time since the big bang lies somewhere between $(2/3) \times (1/H_0)$ and $1/H_0$. The former age arises if the universe contains the maximum density of matter that astronomers accept as reasonable (see Chapter 11), and the latter arises if the universe has a much lower density—so low as to render matter impotent in slowing the expansion. Thus, if astronomers can determine H_0, the value of Hubble's constant now, then the time since the big bang must lie somewhere between $0.67/H_0$ and a time 50 percent larger, namely $1/H_0$. In cosmological circles, to establish bounds to a crucial parameter that describes the universe, and to find that the larger bound exceeds the smaller by only 50 percent, offers cause for congratulation.

So all roads lead to the key question, What is the value of $1/H_0$, the inverse of the Hubble constant today? This might seem an impossible question to answer. The finite speed of light implies that we see other galaxies not as they are but as they were millions or even billions of years ago, when their light left them. But the galaxies whose distances we can hope to estimate are no farther away than a few hundred million light years. A "look-back" time of a few hundred million light years ranks as nearly insignificant in a universe that is many billion years old. Hence we can use these nearby galaxies to determine the value of H_0, the value of the Hubble constant now, without introducing any significant error into our results.

If only all questions related to H_0 could be solved so easily, cosmologists would lie in clover. But as a matter of uneasy fact, determinations of H_0 must confront a hard rock of difficulty: To measure the distance to other galaxies takes decades of work, as we saw in the last chapter, and even then your colleagues may disagree with you.

The Virgo Cluster, the Key to the Hubble Constant

Before we plunge further into the problems of estimating galaxies' distances, let us pause to rejoice that the other half of the Hubble-law equation, galaxies' recession velocities, can be measured with impressive accuracy, often to one percent or even better. Any significant uncertainty on that side of the equation arises not from measuring the velocities but from interpreting those measurements: How much of the velocity arises not from the expansion of the universe but from the galaxies' individual, random motions? As we have seen, this question becomes less important as we look to galaxies with greater distances and thus with larger recession velocities. Unfortunately, progressively larger distances are correspondingly more difficult to estimate, since the galaxies, and any objects within them, appear smaller and fainter.

The ideal galaxies would be those sufficiently close to

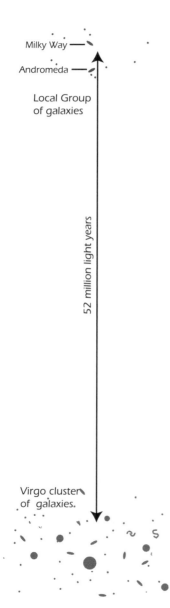

Milky Way

Andromeda

Local Group
of galaxies

52 million light years

Virgo cluster
of galaxies.

us that astronomers can see the details that reveal the galaxies' distances, yet sufficiently far away that their recession velocities arise almost entirely from the universe's expansion. Although such perfection eludes us in the real world, astronomers have long known of, and dreamt of measuring, the galaxies that come closest to the ideal: those in the Virgo cluster.

The Virgo cluster, the closest large cluster of galaxies to the Milky Way, contains about a thousand galaxies and lies in the direction of the constellation Virgo, sprawling over its borders into other constellations nearby. (Of course, these constellations consist of stars in our own galaxy, with distances measured in hundreds or thousands of light years, while the galaxies in the Virgo cluster have distances measured in tens of millions of light years.) This galaxy cluster lies sufficiently far from us that the velocities measured for its member galaxies arise primarily from the expansion of the universe, though individual galaxies' velocities may from the cluster average by plus or minus 5 to 10 percent, thanks to the random component of their motions.

Just how distant is the Virgo cluster? If we knew the answer, we could divide this distance by the recession velocities measured for its member galaxies, which average to about 1,200 kilometers per second, and obtain the value of $1/H_0$. However, we first must adjust for the fact that the Virgo cluster has so much mass that it attracts the Milky Way, with the result that our galaxy is "falling" toward the Virgo cluster. Hence the Virgo cluster's recession velocity, which we measure with respect to the Milky Way, falls below the value we would obtain if we were not falling toward Virgo. This correction amounts to perhaps 10 percent of the measured recession velocity; the fact that we do not know its amount exactly furnishes one of several sources of potential error in the value that we shall derive for $1/H_0$.

If we knew how the distances of galaxies still farther from us, we could divide those galaxies' recession velocities by their distances to find $1/H_0$. (Furthermore, these velocities would be "uncontaminated" by any tendency of the Milky Way to fall toward them.) Because astronomers observe similar types of objects in large galaxy clusters—giant spiral and elliptical galaxies, along with occasional supernovae in those galaxies—they can rather easily obtain the *ratio* of those clusters' distances to the Virgo cluster distance. All they need do is measure the maximum apparent brightnesses of, for example, a supernova explosion in the Coma cluster and one in the Virgo cluster with the same spectral features. If the Virgo supernova appears 36 times brighter, the Coma supernova must be 6 times farther away. What has been missing is an absolute distance to the Virgo cluster, the closest cluster to the Milky Way that can be called representative of the large galaxy clusters seen throughout the universe.

To deal with the actual numbers, suppose that astronomers were to determine that the Virgo galaxies have an average distance of 60 million light years. Then $1/H_0$ would equal the ratio of 60 million light years to 1,200 kilometers per second, the cluster's measured recession velocity. Then $1/H_0$ would equal 60/1,200, or 0.05 second × (one million light years/kilometer).

Say what? The problem is that astronomers use one unit of distance (light years) to describe faraway galaxies, and another (kilometers) when they measure recession velocities. For consistency, they might better measure distances in kilometers, or recession velocities in light years per second. But a foolish consistency is the hobgoblin of little minds, as Emerson said. Better to use the units appropriate to a task, and then make a simple conversion for other purposes. Once we recognize that each light year contains 9.46×10^{12} kilometers, we can see that the value of $1/H_0$ that we obtained above equals $0.05 \times 9.46 \times 10^{18}$, or 4.73×10^{17} seconds.

There you have it: This is the time since the big bang if the rate of expansion has not changed, and if the distance to the Virgo cluster equals 60 million light years. Since each year contains 3.15×10^7 seconds, this time corresponds to 1.5×10^{10} or 15 billion years. Two-thirds of this number equals just 10 billion years, so if the Virgo cluster has a distance of 60 million light years, between 10 and 15 billion years have passed since the big bang.

If $1/H_0$ equals 15 billion years, then H_0 itself equals 1/(15 billion years). In the complex units that we have just examined, this also equals about 20 kilometers per second per million light years. Incredible though it may seem, astronomers often quote H_0 in units of kilometers per second per million light years. They also often measure distances in "megaparsecs," each of which equals 3.26 million light years. Hence, if you associate with cosmologists, you will hear them refer to a Hubble constant, H_0, of, say, 50 kilometers per second per megaparsec, which they abbreviate as "50" because they know the units by heart. In this discussion, I have made the obvious simplifications of using $1/H_0$ as the basic unit of time to date the universe, and of citing values for $1/H_0$ in years rather than in millions of light years per kilometer per second. If you want to keep track of cosmologists' jargon, be aware that if H_0 equals 50 in units of kilometers per second per megaparsec, then $1/H_0$ equals 19.6 billion years; if H_0 is 75 in these units, $1/H_0$ is 13.1 billion years; and if H_0 equals 100, then $1/H_0$ equals 9.8 billion years.

In early 1994, before new estimates of the distance to the Virgo cluster gave cosmologists a spin, the best estimate of the Virgo cluster's distance was about 62 million light years, close enough to 60 million for the 10- to 15-billion-year value that we have derived for the age of the universe to apply (the actual numbers were 10.3 to 15.5 billion). The lower end of this age range seemed unlikely, since astronomers felt confident that some stars have ages of 14 to 16 billion years. But since no distances to galaxy clusters were known to better than 10 to 20 percent accuracy, the upper bound on the age of the universe, allowing for possible errors of this size, gave the cosmos a few billion years' leeway between the time after the universe began and the time when the oldest stars must have formed.

1994, the Year of New Observations

And then came 1994, a year of triumph and confusion, billed by the media (at least those parts that pay attention to the universe beyond Earth) as a "crisis in cosmology." But just as the once-trumpeted Amityville Horror turned out to be closer to the Amityville Nuisance, the assertions of a crisis in cosmology go beyond the actual state of affairs—unless we use the word "crisis" to mean "opportunity for hard work rethinking just what we know."

In the fall of 1994, two independent teams of researchers announced the results of the different methods they had used to obtain better estimates of the distance to the galaxies in the Virgo cluster. Both teams found approximately the same result: The Virgo cluster lies not 62 million but rather 52 million light years from the Milky Way!

The implications of this result deserve our attention before we consider how the astronomers obtained it. If the Virgo cluster is only 52 million light years away, then $1/H_0$, as derived from the Virgo cluster's distance and recession velocity, equals 12.3 billion years. Two-thirds of that value equals 8.2 billion years, so the time since the big bang becomes roughly 8 to 13 billion years. But because models of the universe that favor a high density of matter have become most cosmologists' favorites (as we shall see in Chapter 10), the low rather than the high end of the age range seems favored—at least by those who support these models. In that case, how can a universe that is only 8 or 9 billion years old contain stars that are 12 or 14 or possibly 16 billion years old?

The answer is that it can't. Cosmology poses many conundrums, but it has not yet reached the point of asserting that the universe is younger than its constituents. Clearly modifications must be made in our current picture of the universe, if the new distance to the Virgo cluster proves correct upon further study (as most astronomers expect it will). Astronomers who specialize in estimating the ages of star clusters will have to admit that their estimates of stars' ages are wrong, if the universe proves younger than 10 or even 12 billion years. To the extent that they can defend their models, they will not easily give in. However, as we discuss on page 102, fudge factors give these models some leeway—not quite sufficient to satisfy everyone, but enough to imply that the models may yet prove susceptible to modifications that save the cosmos from being younger than the stars.

Once it was the other way around. The first great "crisis" concerning a discrepancy between the age of the universe and the age of stars rose during the 1930s, when Edwin Hubble and Milton Humason estimated $1/H_0$ (and therefore the age of the universe) as 2 to 3 billion years. Since the solar system itself had already been assigned an age greater than 4 billion years (our current best estimate is 4.6 billion), this result was insupportable by right-thinking astronomers, not to mention the public at large.

The crisis was resolved during World War II, when Walter Baade, a German refugee astronomer who was not allowed to perform military research and who therefore had abundant observing time with the 100-inch telescope on Mount Wilson, profited from the dark skies of the wartime blackout of Los Angeles to take superior photographs of nearby galaxies. From these photographs, Baade discovered that Cepheid variable stars come in two types, and that Hubble had in essence observed one type in the Milky Way and another in nearby galaxies. Not surprisingly, comparison of the stars as if they all belonged to the same class introduced error, in this case by a factor of approximately three.

With this discovery, Baade tripled the estimates of galaxies' distances, reduced H_0 by a factor of three, and tripled the value of $1/H_0$. The universe became at least 8 billion years old, and therefore safe for the solar system and for most stars whose ages were then known. Later estimates raised the value of $1/H_0$ to 15 billion years, and some cosmologists have gone as high as 20 billion. But the recent results have reversed a historical trend by lowering the derived value of $1//H_0$. What are the details behind the distance estimates that have done this deed?

The two new distances derived for the Virgo cluster announced in 1994 used the basic method of observing Cepheid variable stars and comparing them with Cepheid variables in the Milky Way whose distances are known. But the galaxies in the Virgo cluster are at least twice as far from us as any galaxy in which individual Cepheids had been detected previously. The Cepheids in Virgo cluster galaxies have apparent brightnesses so small that detecting them lies just at, or even beyond, the limits of our finest instruments.

Our ability to detect any faint object suffers from the twinkling of the object's light, that is, from the changing amount of refraction that the light undergoes as it passes through Earth's atmosphere. Atmospheric refraction spreads what would be a single, relatively bright point of light into a disk. No single point of this disk has sufficient brightness to be detected against the faint background of light produced both in the atmosphere, from atoms and ions that are excited by fast-moving particles and emit what scientists call the "airglow," and in solar system, where dust particles reflect sunlight to produce the "zodiacal light."

To succeed where previous efforts had failed, the teams of astronomers found two ways to lower the threshold of detectability of Cepheids with low apparent brightnesses. One team, led by Michael Pierce of the Kitt Peak National Observatory, took new, more sensitive detectors of light and attached them to the Canada-France-Hawaii telescope (CFHT) at the Mauna Kea Observatory. At an altitude close to 14,000 feet, sited on a mountaintop that usually stands above all the clouds, this observatory has the clearest skies of any major installation. As a result, the telescopes on Mauna Kea profit from what astronomers call the world's finest "seeing"—nights with the smallest amount of the refraction that makes stars twinkle. In addition, the CFHT has "adaptive optics," which constantly monitor the changing amounts of refraction and move small mirrors that compensate for it, on time scales as short as 1/100 of a second.

With these three advantages—a better detector, an improved telescope, and the best site for it—Pierce and his collaborators found three Cepheid variables in the galaxy NGC 4571, located close to the center of the Virgo Cluster. Repeated observations revealed that the periods of light variation of the three Cepheids lie between 50 and 100 days. By comparing the apparent brightness of each of these stars with those of Cepheid variables with the same period in our own galaxy, the astronomers derived a distance to NGC 4571 of 48.6 million light years, with a possible error that they estimated to be plus or minus 3.9 million light years.

Going farther than the Mauna Kea Observatory, the other team of astronomers used the Hubble Space Telescope (HST) for a similar endeavor. Orbiting above the entire atmosphere, the HST offers astronomers a twinklefree view of the cosmos, and a correspondingly increased ability to detect faint objects. When the HST was first conceived, early in the 1970s, one of its key missions was seen to be exactly what these astronomers achieved: the identification of Cepheid variable stars in galaxies belonging to the Virgo cluster. The HST astronomers, led by Wendy Freedman of the Carnegie Observatories, found more than a dozen Cepheid variable stars in M100, another member of the Virgo Cluster (see Figure 11), with periods of light variation that range from 22 to 53 days. By comparing the apparent brightnesses of these Cepheids with those of similar stars in the Large Magellanic Cloud, whose distance we think we know to within a few percent, Freedman and her collaborators found a distance to M100 of 55.7 million light years, with a possible error of plus or minus 5.9 million light years.

Of course, both groups of astronomers recognized that these distance determinations contain many potential sources of error, including imperfections in the detectors that measure stars' apparent brightnesses, possible mistakes that could arise in the comparison of Cepheid variables or in determining the distances to the closer Cepheids, and the fact that the various galaxies in the Virgo cluster do not have precisely the same distances. These problems are what led the two groups to provide their distance estimates as a number plus or minus 4 to 6 million light years, that is, plus or minus about 10 percent. The uncertainty in how much of the galaxies' recession velocities arises from the expansion of the universe increases this potential error to about plus or minus 20 percent. Thus, taken all in all, the new results of 1994 for the Virgo cluster's distance imply that $1/H_0$ equals 11.8 billion years, with a possible error of 20 percent or so in either direction.

The two estimates of distances for galaxies in the Virgo

cluster, made independently though announced nearly simultaneously, tend to reinforce one another, because the plus-or-minus limits from both groups include values close to 52 million light years. Furthermore, the two groups of astronomers worked on different galaxies, which may well differ by a few million light years in their distances from us.

These distance estimates receive further support from another method used to derive distances, pioneered by the Harvard astronomer Robert Kirshner. This method relies on observations of exploding stars—supernovae—which can be recognized out to vast distances from the Milky Way. The supernovae used to estimate distances come in two types. One of these, called *Type Ia supernovae,* arise when a white dwarf star forms part of a binary star system, and has a companion star from which the white dwarf acquires hydrogen-rich material. Once a sufficient amount of matter accumulates on the white dwarf's surface, a thermonuclear explosion can occur. Astronomers now think that they can determine the luminosity of these Type Ia supernovae to an accuracy of 10 percent, so that supernovae of Type Ia can serve as standard candles, consistent to 10 to 20 percent in their luminosities.

The second way to employ supernovae as distance markers uses astronomers' understanding of how some stars explode. As we described in Chapter 3, the core of a massive star will collapse when the star exhausts all further possibilities for nuclear fusion in its interior. This collapse produces a shock wave that explodes outward from the center of the star, shooting the star's outer layers into space as a *Type II supernova.* In studying the spectrum of light emitted by a

Type II supernova, astronomers can use the Doppler effect to find the speed at which the light-emitting gas blasts outward. They can also determine the temperature of these fast-moving gas layers. Finally, supernova experts know how to calculate the luminosity of a spherical shell of gas of a particular radius and temperature.

These data allow astronomers to estimate the distances to supernovae. They can calculate the increase in the radius of the supernova's outer layers over any interval of time, because this increase equals the time interval multiplied by the expansion velocity, which is revealed by the Doppler effect. Furthermore, the ratio of the supernova's radius to its distance can be directly related to the supernova's apparent brightness and temperature, so observing the latter two quantities provides that ratio. Thus, with a knowledge of the supernova's temperature, its apparent brightness, and its change in radius with time, astronomers can derive the distances to a Type II supernova.

Thus Type Ia and Type II supernovae offer two separate ways to estimate distances and thus to determine the Hubble constant. In 1994, Robert Kirshner and his colleagues applied both methods to the determination of the distances to galaxies in which supernovae appeared. They then used these distances, along with measurements of the galaxies' recession velocities, to redetermine the Hubble constant. Use of the Type Ia supernovae as distance indicators gave a value for $1/H_0$ of 14.6 billion years, while the use of Type II supernovae led to a value for $1/H_0$ of 13.4 billion years. Both of these values have an error of about 10 to 15 percent.

If we take the tempting course of averaging the two supernova-based results, the resulting value of $1/H_0$ equals

14.0 billion years, and (2/3) H_0 equals 9.3 billion years. The supernova-based distances leave the universe about a billion years older than the Cepheid-determined distances do. If improvements in the supernova distances verify that $1/H_0$ is close to 14 billion years, a problem will still exist in matching the ages of the oldest stars to the age of the universe—but the problem will be a billion years smaller.

Taken together, the four separate results for $1/H_0$ announced in 1994 imply a value close to 12–13 billion years. Although the value of H_0 cannot yet be regarded as certain, the new results obtained from the Cepheid variables in two galaxies in the Virgo cluster and from supernovae in even more distant galaxies do overlap; few astronomers doubt that the result could be mistaken by more than, say, 30 percent at the most. One who doubts this, however, is Allan Sandage, the heir to Hubble's work at the Carnegie Observatories. Sandage believes that the new results are replete with errors, and that $1/H_0$ is greater than 16 billion years. But if we accept the 1994 observations, we can state that the time since the big bang now seems to lie somewhere between 8 and 14 billion years—less than the ages currently estimated for the oldest stars. And so as Lenin asked before he fell into the dustbin of history, What is to be done?

A conservative might say that since the estimated error in the determination of H_0 amounts to at least plus or minus 15 percent (and might be as large as 20–25 percent), then if we take 15 percent upward from 13 billion years, so that $1/H_0$ equals 15 billion years, and if we assume that the density of matter in the universe is low, we really don't have a problem, since nearly all stars' ages can fit into this time frame. But we should save this sort of special pleading until we are completely backed against the wall of time. For now, we may summarize the recent determinations of the age of the universe as follows:

Many cosmologists now believe that the total density of matter in the universe has a sufficiently high density that the time since the big bang can be no greater than $0.8/H_0$ (see page 139). (Recall that only an extremely low-density universe has an age of $1/H_0$.) The new determinations made in 1994 set the value of $0.8/H_0$ between 9.4 and 11.2 billion years, with a central value of 10.3 billion years, and an error estimated at plus or minus 20 percent. The time since the big bang therefore equals no more than 7.5 to 12.5 billion years, which we shall refer to henceforward as 8 to 13 billion years without much loss of accuracy. If the density of matter turns out to be only half as great as the current minimum estimate, this age range would increase to $0.9/H_0$, and thus would span the interval of 8.4 to 14.0 billion years. On the other hand, if the density of matter were equal to the maximum value that theoreticians consider, the so-called "critical density" that is predicted by the popular "inflationary model" of the universe described in Chapter 10, then the time since the big bang would lie between 6.3 and 10.5 billion years, times so small as to be quite troubling. Even the greater age of 8 to 13 billion years for a less dense universe still raises the question, What is to be done?

A Nonzero Cosmological Constant

The more radical answer to What is to be done? is to take the cosmological constant more seriously. A little history is in order.

As we have seen, when Einstein and other cosmologists

pondered the general theory of relativity, they saw that it implies only two cosmic possibilities: expansion or contraction. No "static" universe, which neither expands nor contracts, can ever exist. Because Einstein felt that this was unacceptable, he modified his equations to include an additional term which came to be called the cosmological constant. The cosmological constant—if it exists—gives empty space certain counter-intuitive properties: it has positive energy but exerts negative pressure, equivalent to a tension, as if space were being squeezed and twisted—by space! In this case, every empty cubic centimeter in space contains a constant amount of energy. Thus, as the universe expands, it creates not only new space but also new energy.

The negative pressure of empty space means that space automatically tends to expand, which opposes its tendency to contract from the mutual gravitational attraction of each part of the universe for all the other parts. In this way, with the cosmological constant set at the appropriate value, the universe can be static after all.

By introducing the cosmological constant, Einstein was quite understandably attempting to produce a theory that would describe the real universe, rather than an imaginary one that might be describable in a consistent way but which would turn out to exist only on paper. After he learned from Hubble's observations that the real universe is not static but is expanding, he saw no further use for the cosmological constant. Years later Einstein commented to George Gamow that its creation was "the greatest blunder of my life."

In making this assessment, Einstein followed a golden rule of science: When evaluating competing explanations for a set of facts known from observation to be true, choose the simplest explanation. This rule sometimes goes by the name of "Occam's razor," after the fourteenth-century English clergyman and philosopher William of Ockham (for historical reasons, the man's town and his razor have acquired different spellings). Occam is credited with having written that "entities are not to be multiplied without necessity," though what he wrote was apparently more like, "It is vain to do with more what can be done with fewer," which expresses the idea nearly as well. Occam's conclusion has made good sense to scientists since then, who have used it to pare away what they judge to be needless hypotheses—though witty and insightful cosmologists have on occasion wondered whether God must shave with Occam's razor.

Today, seven decades after Einstein first introduced the cosmological constant, it lives on in the hearts and equations of many cosmologists, in large part because it offers a way to explain details of the universe unknown to Einstein in the 1920s. Even though the universe is indeed expanding, so that the original motivation has fallen away, a cosmological constant might well exist, though it need not have exactly the value that would keep the universe perfectly static. Instead, with a nonzero cosmological constant, instead of slowing in a regular manner, the expansion of the universe could first slow significantly, then "coast"—that is, continue at a steady rate—and then slow down some more. When cosmologists work out what this means in terms of H_0, they find that the universe could be *older* than $1/H_0$ and still appear as it does.

This outcome would prove extremely handy in reconciling the real universe with our models if the time since the big bang turns out to be close to 8 billion years in models with a cosmological constant equal to zero. If we are forced to conclude that calculations of stars' ages cannot be reconciled with the best estimates of the age of the universe, Einstein's "greatest blunder" could allow us to reconcile theory and observation.

The cosmological constant provides a near perfect fudge factor, because some value of the cosmological constant will brilliantly reconcile the ages of the universe and of the stars, and we have few other constraints on the value of this term in the equations. This situation may not last for long: A large value for the cosmological constant would imply many more cases of "gravitational lensing"—the amplification of light from a distant galaxy or quasar as it passes close by a massive object—than are now observed. Further observations will refine this statement still further.

The spirit of Occam's razor cautions us not to adopt a nonzero cosmological constant until that becomes absolutely necessary. Remember that the cosmological constant may have any value, say from 10^{-40} to 10^{40} in units that need not concern us here. Most cosmologists therefore find it passing strange for it to have just the value we require to make our model fit the real universe. For now, most cosmologists would say that the strongest argument against a nonzero cosmological constant consists of the fact that we have no way to explain why, amid the enormous range of possibilities, the cosmological constant would have the one particular value that explains the universe. The only aesthetically appealing value for the cosmological constant is zero—but the day when we can determine the degree to which our aesthetic rules the universe, at least in terms of the cosmological constant, has not arrived.

Could the Star Ages Be Incorrect?

And when will that day come? Only when we have more accurate observations. For the time being, the results of 1994 make certain tasks clear. Astronomers who specialize in distance determinations must continue to check their recent derivations, while other astronomers, experts in estimating the ages of stars, must rethink their results of long standing.

As we saw in Chapter 4, astronomers' estimates of the ages of the oldest globular star clusters, 14–16 billion years, derive from computer-calculated models of nuclear fusion within stars and how that fusion changes as the star grows older. Astronomers use computers because the models are complex; they must specify how the star transports energy outward through every point in its interior. For forty years, astrophysicists who specialize in stellar interiors have refined their calculations of the structure and evolution of the stars and have applied their results to derive the ages of stars in clusters. Some experts, such as Pierre Demarque at Yale University, consider that the ages calculated for the stars in the oldest globular clusters, the 14–16 billion years cited above, cannot be significantly reduced without abandoning our claims to understand what happens inside stars. Others, however, including David Schramm of the University of Chicago, have concluded upon reexamination of the computer models that some fudge factors do exist. Schramm's models suggest that if these fudge factors all run in the direction of lower ages for the stars, then the upper bound on age becomes 12 billion years, plus or minus the 2 billion years that represents possible errors in the calculations.

If Schramm's models, or something like them, turn out to be correct, and to have their fudge factors pointing in the right direction, then the oldest globular clusters may have ages of "only" 10 to 12 billion years. In that case, only the inflationary model, which requires that the age of the universe equal $0.67/H_0$, would be in danger of exclusion as contradictory to the ages of the stars, since the 1994 results imply that $0.67/H_0$ equals 6.3 to 10.5 billion years. And even the inflationary model, which Schramm and many other cosmologists still favor, could survive if the fudge factors all fall into the proper place.

In brief, then, the modern "crisis in cosmology" could well pass into history without leaving behind even the impact that was produced 50 years earlier by Walter Baade's resolution of the "age crisis." On the other hand, if further observations show that the relatively low values for $1/H_0$ are correct, and if further computer models prove that the fudge factors for stellar ages are incorrect, then astronomers will indeed face a crisis. They will certainly surmount it, survive it, and continue to contemplate the future of the cosmos along with the past that brought us here.

NINE

An Uncertain Future

Cosmologists have their areas of uncertainty, but they feel confident that the future of the universe must follow one of two courses: either eternal expansion, or reversal into contraction at some point in the distant future. Realms of philosophy lie wrapped within these two possibilities. Eternal expansion will produce a cosmos of ever-diminishing expectations, a cold, dark universe with a density so low that it makes the present universe seem stuffed with matter. Eventual contraction, far more attractive emotionally to many, will finally—far into the future—bring a universe that ceases its expansion and reverses itself into a contracting phase. Calculations of how the expansion rate has changed since the big bang imply that a contraction phase, if it occurs at all, lies at best a hundred billion years in the future. The logical outcome of the contracting phase would be a "big crunch," in which all space and matter occupies a volume infinitesimal in size.

Such a big crunch might be followed almost immediately by another big bang, though as yet we have no good mathematical theory that calls for such a result. But the alternatives are clear: The universe is either a one-shot affair, endlessly growing more vacuous, or a composite of expansion and contraction, offering the knowledge that all our errors and our garbage will eventually be recycled through the big compactor that lies at least a hundred billion years in the future (see Diagram 6).

This time represents a lower bound, and in fact we do not know that the universe will ever contract in a big crunch. If the universe expands without end, then the glorious structures that the universe has produced may well persist, even as the stars that reveal them fade into blackness. On the other hand, the contraction of the universe would destroy every structure and indeed every trace that such structure ever existed.

The Importance of Density

The future of the cosmos does not depend on human emotions or philosophy. The utterly dominant factor, so far as the future of expansion is concerned, is a single parameter: the average density of matter in the universe. Density measures the amount of mass per unit volume. Mass plays a crucial role in determining the amount of gravitational force. As Newton showed, the strength of gravity between any two regions of space varies in proportion to the product of the mass contained in each region, divided by the square of the distance that separates them.

If the destiny of the universe includes an end to its current expansion, this will arise because of gravity. Gravitational forces attract all objects with mass toward all other objects; gravity therefore brakes the expansion, opposing the now-prevailing tendency of the universe to expand that was established in the big bang. If we like, we can imagine that the big bang imparted a tremendous kick to all the matter—and all the space!—in the universe. Since then, all regions of space have expanded away from all other regions, but they have not done so in an entirely constant manner. Instead, because gravity is the only force that acts significantly over distances of billions of light years, the expansion of the universe has been slowing down ever since the moment of the big bang.

Of course, slowing down does not necessarily mean ever reaching zero. If we launch a rocket upward, Earth's gravity will surely tend to *slow* the rocket, but the rocket may nevertheless coast away from Earth indefinitely. Similarly, the universe today, many billions of years after the big bang, goes right on expanding. It does so at a rate that will double the distance between any two representative clusters of galaxies after another 8 to 13 billion years have elapsed. But this rate is lower than the rate of expansion in bygone eras. When the universe had half its present age, for example, the rate of expansion was such that distances would double after another 4 to 6 billion years.

Roughly speaking, the universe will double all distances in a time approximately equal to the time that has elapsed since the big bang. This matching of the amount of time since the big bang with the time required for cosmic distances to double does not arise by coincidence. Instead, the match comes straight from modern understanding of the expanding universe. With a look at Einstein's general theory of relativity, we can gain the basics of this understanding.

The Curvature of Space

Einstein's general relativity theory describes how gravity affects space. Fundamentally, gravity bends space, and the bending of space tells matter how to move. Greater concentrations of mass produce more gravitational force, and hence greater bending of space, than lesser ones do. (For reasons of both time and space, we must omit a discussion here of what it means to say that space *has* a shape—especially if space has no existence independent of matter. From time to time, we must make such sacrifices to stay afloat in the cosmological sea.)

Unlike the rest of us, cosmologists (no doubt through long practice) experience no difficulty in imagining curved space. If space has positive curvature, it curves back on itself in a manner analogous to the surface of a sphere (see Diagram 3). Einstein's theory of general relativity shows that a universe with positive curvature must eventually contract. In contrast, a universe with negative curvature will expand forever. Space in such a universe may be analogized to a saddle. However, since the key result of a negative curvature of space—eternal expansion—is the same as that for a universe with zero space curvature, we can simply deal with zero curvature as leading to the opposite of eventual contraction.

Flat space—space with zero curvature—extends to infinity in all directions, just as our intuition insists space must, and negatively curved space likewise has an infinite extent.

The possibility of curved space forms the heart of Einstein's theory of general relativity, and the theory leads to an equation that describes the general evolution of the universe, which we may call the *evolution equation*. The evolution equation can be posed in two ways, depending on the side of the equation to which we move appropriate terms.

If the cosmological constant is zero, the evolution equation that describes the ongoing evolution of the universe is $H^2 = (8\pi G\rho/3) - (\kappa\, c^2/S^2)$, where H is Hubble's constant, which, as we have seen, appears in Hubble's law to express the proportionality between galaxies' distances and recession velocities. On the right-hand side of the equation, G is the gravitational constant, ρ is the actual density of matter, κ is the (constant) curvature of space, c is the speed of light, and S is a scale factor that measures the average separation of two representative points such as galaxy clusters.

The average density of matter in the universe ρ (mass per unit volume) varies in proportion to one over the cube of the scale factors. As the universe expands, $(8\pi G\rho/3)$, which is proportional to ρ and thus to $1/S^3$, *decreases more rapidly than the second term*, which is proportional to $1/S^2$. This is not important if κ is negative, because then both terms on the right-hand side of the equation make a positive contribution. H^2 will therefore always be greater than zero, and the universe will expand forever. If κ is zero, H^2 will fall to zero only when the universe becomes infinitely large, and the average density of matter then decreases to zero. But if κ is positive, H^2 may be positive at some relatively early time in the universe—because the first term in the equation is greater than the second—but will eventually decrease all the way to zero as the universe expands and S grows larger. In other words, the expansion will never cease in a negatively curved universe, will cease only after an infinite time in a flat universe, but will stop at some finite time in a positively curved universe (see Diagram 6).

The second way to write the evolution equation employs the concept of the critical density, ρ_c, which is defined as the ratio $3H^2/8\pi G$. If we multiply both sides of the evolution equation by $(3/8\pi G)$, substitute for ρ_c, and rearrange the terms, we obtain the equation $\rho - \rho_c = (3\kappa c^2/8\pi GS^2)$. This shows that if the actual density of matter is greater than the critical density, then the curvature of space, κ, must be positive, whereas if the actual density is less than the critical density, the curvature of space must be negative.

It is this fact that makes the critical density useful, and indeed inspires its definition: Whether or not ρ exceeds ρ_c is equivalent to whether or not κ is positive or negative. Note that since the curvature of space does not change (nor do c or G), and since S is always positive, the difference between ρ and ρ_c, $(3\kappa c^2/8\pi GS^2)$, can never change sign, and must always be positive, always be negative, or always be zero (except for the case where we wait an infinite amount of time in an expanding universe, so that ρ finally decreases to equal ρ_c, but only after an infinite amount of time has passed). Cosmologists write the ratio of ρ to ρ_c as Ω (the capital Greek omega). If omega exceeds 1, the universe will eventually contract; if omega is less than or equal to unity, the universe will expand forever.

This second way to write the cosmic evolution equation emphasizes the importance of the critical density. If space has positive curvature, the actual density of matter must be *larger* than the critical density. Likewise, a negative curvature of space implies that the actual density must be *less* than the critical density; a zero curvature of space implies that the average density must *equal* the critical density.

Of course, we know that the average density of matter decreases continuously as the universe expands: The same amount of matter fills a larger volume of space. But as the universe expands, the value of the critical density falls too, and in the same proportion as the actual density. The evolution equation shows that if one density exceeds the other, it will do so for any finite amount of time into the future. Therefore, if the actual density exceeds the critical density *now,* it has always done so and will always do so. Likewise, a value of the actual density below the critical density implies that the real density always has been, and always will be, less than the critical density.

Both the future of universal expansion and the difference between the actual and critical densities depend on the curvature of space. The evolution equation therefore implies that if we can determine how the actual density of matter compares with the critical density, we can thereby discover the future of the universe. If the average density of matter is less than the critical density, the universe will expand forever. And if the actual density of matter should exactly equal the critical density, then the universe will stop expanding—

but only after an infinite amount of time has elapsed. This amounts to saying that an average density exactly equal to the critical density implies a universe whose expansion rate heads toward zero but never quite gets there in any finite amount of time.

The critical density is a number that we can calculate, provided that we know the value of Hubble's constant. Once we have made this calculation, we can compare the critical density with the actual density of matter in the universe. If the latter exceeds the former, the universe will eventually contract; otherwise, it will expand forever. For six decades, a great goal of observational cosmology has therefore been to determine the average density of matter in the universe, either by direct measurement (extremely difficult, since we cannot travel through, say, a million cubic light years to sample the matter it contains) or by more roundabout methods of deduction. If we can find this number, we can immediately compare it with the critical density to determine whether or not the universe will expand forever.

Today, billions of years after the big bang, the critical density equals approximately 1.2×10^{-29} gram per cubic centimeter. There is an uncertainty in this number that arises from our uncertainty about the exact value of H_0, whose inverse gives the time needed now for representative distances in the universe to double. This uncertainty in turn arises from our difficulties in measuring distances—in particular, the distances to the galaxy clusters whose recession velocities provide the basis for concluding that Hubble's law is valid. As we saw in the last chapter, the distances are uncertain to plus or minus 20 percent, but the uncertainty in the value of the critical density varies in proportion to the square of the uncertainty in the distances. Since 1.2 squared equals 1.44, it is fair to conclude that the value of the critical density is uncertain by about 40 percent: It might be as large as 1.7×10^{-29} gram per cubic centimeter, or as small as 0.8×10^{-29}.

Either way, as Pogo said in a slightly different context, it's a mighty sobering thought. A density of 1.2×10^{-29} gram per cubic centimeter is about equal to what you get by placing 7 atoms of hydrogen in a cube one meter on a side. Surely the universe must have a density many times greater than this tiny amount—so that eventual contraction of the universe is assured!

Determining the Average Density of Matter in the Universe

Quite possibly not. The most crucial—and almost certain—fact that we know about the average density of matter in the universe today is that all the matter we can see, and indeed all the matter that we may consider to be "ordinary" matter, provides no more than 2 percent of the critical density. Astronomers have derived the first part of the statement from direct observation, more or less. They know how to estimate the distances to galaxies from measurements of their spectra and the use of Hubble's law, and they know approximately how much mass a typical large galaxy contains in its stars. By observing a large number of galaxy clusters, astronomers have found that all the stars in all the galaxies apparently provide a total density less than 0.5 percent of the critical density. A more subtle approach, described on page 171, sets the total density of ordinary matter at about 4 to 8 percent of the critical density.

Therefore, if we must rely upon the matter that shines in stars, in galaxies, in everything that lights up the universe with its glorious forms and colors, then we are doomed in our hopes of eventual contraction. We may total all the mass we think that giant galaxies embrace, each with their hundreds of billions of stars much like the sun; we can throw in smaller, satellite galaxies; we can add supermassive black holes at the centers of galaxies, each with a billion times the mass of the sun; but when we divide the total mass within a representative volume by the amount of space within that volume, we still fall far short of the critical density. If we hope for the eventual contraction of the universe, we must hypothesize that the universe contains many times more matter of some totally unobserved, unknown form than it does of the ordinary matter that we have identified.

Why should we ever expect that the universe contains far more nonluminous matter of unknown form than appears in the form of luminous matter? One part of the answer is startling. As we shall see in more detail in Chapter 11, astronomical measurements of the motions of stars in our own Milky Way galaxy, and of individual galaxies in galaxy pairs and galaxy clusters, imply that the universe contains at least 5 to 10 times more mass than the luminous matter provides. This by itself is a remarkable fact, but a still more startling possibility is this: According to the most widely accepted theoretical model of the earliest moments of the big bang (see Chapter 10), the density of the universe is, of necessity, equal to critical density, and therefore the mass that is currently "missing"—the mass that exists but hasn't yet been detected or identified—provides for more mass than the matter we see and are familiar with.

What could this "missing mass" or "dark matter," as cosmologists call it, possibly be? One key fact to bear in mind is that the dark matter almost certainly consists of "strange matter"—matter unlike anything we know now. If the nonluminous matter were something like familiar matter and were much more abundant than the matter we see, then it would have left behind a fingerprint. Like the dog that did not bark in the night in the Sherlock Holmes story "Silver Blaze," the absence of that fingerprint implies that this dark matter must be strange indeed.

This missing fingerprint resides in the relative abundances of the isotopes of hydrogen and helium, which we can measure today. These abundances were established by nuclear fusion in the early universe; and their values depend on the density of ordinary matter in the universe soon after the big bang. As discussed in more detail in Chapter 12, the measured abundances of hydrogen and helium isotopes do *not* reveal the fingerprint we would expect if the density of ordinary matter were even close to the critical density. This fact rules out ordinary matter as the dark matter, no matter how little light ordinary matter may emit.

As a result, few astronomers were surprised in 1994 by the announcement that observations made with the Hubble Space Telescope (HST) had failed to reveal sufficient numbers of faint, cool stars, called red dwarfs, for red dwarfs to qualify as the dark matter. A much greater shock would instead have arisen if the HST *had* found red dwarfs in the enormous numbers required. To most astronomers, this hardly seemed possible, since red dwarfs consist of ordinary matter, which must inevitably have left its fingerprint on the abundances of the isotopes of the lightest elements.

A second key point about the dark matter is that even if it dominates the cosmos, this does *not* mean that the average density of matter in the universe must equal the critical density, or even come close to it. Observational evidence for the existence of dark matter (which we shall encounter in Chapter 11) implies the existence of about 5 to 10 times more dark matter than visible matter. Since the density of visible matter amounts to only about half a percent of the critical density, adding the dark matter raises the average density in the universe to something like 5 percent of the critical density. Impressive though the dark matter may be in comparison with visible matter, we are still far from the critical density. Of course, one might say that if we can find so much dark matter, why not find 10 or 20 times more, and thus obtain a density greater than the critical density, to assure that the universe will contract?

For now, this argument won't take us far. The observational evidence points toward a total density of about 10 percent of the critical density. The push for a density *equal* to the critical density arises not from observations but from theories of the universe—in particular, from the inflationary theory discussed in Chapter 10. So when one hears the phrase "dark matter," it is important to remember to distinguish between the dark matter revealed by deduction from actual observations of the universe, and the far greater amount of dark matter hypothesized by theorists. The theorists may yet prove to be correct and the missing mass may be found, but today the evidence cannot stretch so far without seriously enlarging the possible errors that observational astronomers assign to their results.

What is this dark matter? And how do we know that it exists, if it doesn't shine? The second question has a reasonably good, though fairly complex answer, but as things stand now we have basically no idea what the dark matter is, only what it is *not*. Chapters 11–14 address the issue of the mysterious dark matter, and the search for its nature, in some detail.

But first let us summarize what we know about the future of the universe. With a cosmological constant of zero, the evolution equation implies a definite relationship between the age of the universe and the time required for representative distances to double. In a low-density universe those numbers must be nearly equal: the distance-doubling time equals the time elapsed since the big bang. If, on the other hand, the universe has a density of matter approximately equal to the critical density, the result is somewhat different: The time for representative distances to double equals about 1.5 times the age of the universe.

In both cases, the key fact is that as the universe ages, the time for distances to double increases in proportion to its age. This is another way to say that as the universe goes on expanding, the *rate* at which it expands, expressed by the Hubble constant H, decreases continuously. Whether the doubling time equals the time interval since the big bang or 1.5 times that interval hardly matters; the crucial point is that the distance-doubling time steadily decreases.

But will it decrease all the way to zero, that is, will the expansion cease entirely? This remains a crucial unanswered question of cosmology, the solution to which depends on the average density of matter in the universe. Before we plunge further down that murky dark-matter road, we must make an excursion into the arena that has provoked the modern quest for critical density, the inflationary model of the universe.

The Inflationary Theory

Of all the strange theories about the universe that have been bandied about by cosmologists and mused about in popular science magazines—superstrings, time reversal, quantum gravity, and a host of others—we may expect that some will turn out to be amazing but almost certainly incorrect descriptions of the universe, while others, just as fantastic, will prove to be true. Of all these candidate theories, one stands out among all others for combining unbelievability with an apparently large likelihood of accuracy: the inflationary universe.

The word "inflation," coined for this cosmic purpose at a time when consumer prices were rising rapidly, refers to a period during the earliest moments of the universe when all of space and everything in it was expanding at a tremendous rate. Though all astronomical facts may seem hard to swallow, few can match the scenario proposed by the inflationary universe: During inflation, the distance between any two representative points doubled, then doubled and doubled again, perhaps 50 or 100 doublings in all, with each doubling taking a time no longer than 10^{-33} second!

To picture the early universe in the inflationary theory, imagine a place of seething, high-energy turmoil, in which any two points that might be close to one another now would find themselves twice as far apart 10^{-33} second later, and four times farther apart after twice that time interval. Like the power of compound interest, the effect of repeated doublings grows to be enormous. Ten doublings correspond to a factor of just over a thousand (1,024, to be precise); twenty doublings amount to more than a millionfold increase (1,048,576, to be picky). With fifty doublings we increase by more than a thousand trillion times (10^{15}), with one hundred doublings, by more than a million trillion trillion (10^{30}), and with a thousand doublings, by more than 10^{300}, a number that would fill several lines with zeroes.

So when cosmologists airily speak of an inflationary phase of the early universe that involved hundreds of thousands of doublings in size, they are talking about a significant increase. If you double the size of an atom 50 times, it becomes the size of Delaware; if you double its size 100 times, the atom grows larger than a cluster of galaxies. Who has the audacity to imagine a universe that expands atom-sized regions of space to galaxy-cluster sizes in 10^{-30} second? What gives them the right to think along these lines? What cosmic forces could possibly make the early universe double in size 50 or 100 times?

The answers to these questions are, first, that cosmologists are not afraid of large numbers; second, that they created the inflationary universe not by accident but because this model appears to explain a great deal about the universe that we observe; and, third, that according to the inflationary model, space itself contained the reason for its enormously rapid inflation. Let us examine the third of these statements to see whether we can crack the mysterious shell of scientific talk to fry the yolk of understanding.

The False Vacuum

The inflationary universe makes no sense without a *mechanism* for the inflation. Although various suggestions have been made for such a mechanism, these mechanisms share the common concept that space may not be as empty as it seems; what appears to be nothing but the vacuum, containing nothing but empty space, could turn out to be what particle physicists call *false vacuum*. During the late 1970s, elementary-particle physicists, most notably Sidney Coleman and Demosthenes Kazanas, had suggested that false vacuum is packed with energy—energy that, many cosmologists believe, made the cosmos expand at a fantastic rate during the inflationary era that began the universe.

If "inflation" appears unreasonable, what can be said of the false vacuum? Our experience with empty space is quite limited; in fact, as believe-it-or-not columns like to emphasize, we cannot create perfectly empty space in this world of ours. Even the best "vacuum" on Earth contains a multitude of gas molecules, so it would be unrealistic to say that we can be sure of the properties of totally empty space by direct observation. Furthermore, the cosmologists who talk of false vacuum do not mean that "real" vacuum—truly empty space—cannot and does not exist. Instead, the theorists hold that the universe may contain enormous stretches of truly empty space (true vacuum), and that it was only during the earliest moments of the universe that what "looked" like space was in fact false vacuum.

False vacuum would appear empty but would in fact teem with energy. There is no question here of hiding the energy in the form of fast-moving but invisible particles, such as photons with wavelengths longer or shorter than those of visible light, or within a black hole. Nor are we discussing types of particles not yet known to our experience, such as those that enter the discussion of dark matter. We must instead consider the possibility that space *itself* possessed energy—the scenario we presented in the previous chapter when we discussed the cosmological constant. And indeed, false vacuum behaves just like empty space in a universe with a particular nonzero cosmological constant. (The cosmological constant can, theoretically, have one of many values, only some of which, however, make empty space behave like a false vacuum.)

Water

In their efforts to explain the false vacuum to a waiting world, the best analogy that scientists have produced invokes the latent heat that appears during a phase transition, as when water freezes into solid ice. Even though the ice has the same temperature as the water, the act of freezing produces latent heat—energy—that must be given up to the environment as the water freezes. Similarly, to make ice melt into water, without changing the temperature from 0 degrees Celsius, one must add to the ice the same amount of latent heat that appears upon freezing.

In this analogy, transparent water represents the false

Energy liberated

Ice

vacuum, with energy that will appear upon freezing, and equally transparent ice represents the real vacuum. But an imperfection in this analogy quickly arises. The false vacuum has a property completely unknown in the realms of ice and water. As the false vacuum expands, each cubic centimeter continues to have the same amount of energy as before. The decrease in the amount of energy per cubic centimeter that anyone would reasonably expect from the expansion simply does not occur. Instead, since the amount of energy per cubic centimeter is a characteristic of the false vacuum, when more space appears, the total amount of energy in all of space rises in direct proportion!

There is no good way to make this appear immediately reasonable to the human mind, but false vacuum might nevertheless have been the way the universe was. The false vacuum has a distinguished pedigree, since it is just empty space if the universe has a nonzero cosmological constant. As we have seen, Einstein invented the cosmological constant to allow the possibility of a static universe, but in the context of modern particle physics, we now see what the cosmological constant means—false vacuum.

In a universe with a significant density of matter, the cosmological constant allows for a long coasting period, during which the universe neither expands nor contracts significantly. This has its own interest for those who delve into the issues of how the universe has evolved, for a long coasting period might solve some of the problems of the formation of galaxies and of still larger structures, as we have seen. But the cosmological constant may play another role, a role of supreme importance during the earliest moments of the universe. For those moments, the existence of the

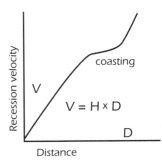

cosmological constant could have produced expansion at an extremely rapid rate—hence the inflation that lies at the heart of the inflationary model of the universe.

According to the inflationary theory, the universe underwent amazing changes during its early history. Immediately after the big bang, we are hard pressed to say just what conditions existed—for example, whether empty space in the universe consisted of false vacuum or true vacuum. Conditions throughout the entire cosmos were then so extreme that we may fairly say that nothing on Earth tells us what we should expect to have been the universal rule soon after the big bang. Furthermore, the rules of quantum mechanics imply that we never can know what was going on in the universe before the Planck time, the first 10^{-43} second after the big bang, had elapsed. Prior to that, the universe was so mixed that space and time could not even be distinguished from each other. Cosmologists currently have no viable theory to deal with these conditions.

But after the Planck time, when more than 10^{-43} second had passed since the big bang, some portion of the universe consisted of false vacuum, with its astounding property of maintaining the same amount of energy per unit volume. This property would have made the tiny "bubble" of false vacuum explode. The rate of expansion at any time depends on the energy density, that is, the amount of energy per unit volume. Since the false vacuum possesses a tremendous amount of energy per cubic centimeter, a bubble of false vacuum will expand at a fantastic rate.

The inflating bubble would have doubled in size every

An Inflationary History of the Universe

10^{-34} second or so. Every ten doublings—which would have required a total of some 10^{-33} second—would inflate the bubble by a factor of 2^{10}, or 1,024. A hundred doublings would inflate the bubble's size by 2^{100} (or about 10^{30}) times, and 200 doublings by a factor of 2^{1000}, or 10^{60}. Since 10^{30} is the factor by which the diameter of the Milky Way galaxy exceeds the size of an atomic nucleus, an increase of size by a factor of 10^{60} would turn any initially subatomic bubble of space into a region far larger than the visible universe we see today.

The third step in the history of the universe, according to the inflationary theory, happened a good deal later, at a time about 10^{-30} second after the big bang. (Notice that when the universe was 10^{-34} second old, at the time that inflation began, a time of 10^{-30} second *was* a long time— 10,000 times longer than the universe had then existed.) By that time, the inflating bubble had doubled in size at least several hundred times. A second phase transition then occurred, in which the false vacuum became true vacuum. This transition released enormous amounts of energy into the true vacuum, causing a "reheating" of the universe, so called because the big bang had heated the cosmos, but this heat had been dissipated during the preinflationary phase. The reheating occurred when the bubble had already grown far larger than the universe we see, and left it made of true vacuum plus the particles that emerged from the energy contained in the false vacuum, and expanding according to Einstein's familiar formulas, with a zero cosmological constant. (When I say zero cosmological constant in this context, I really mean "the same cosmological constant that the universe must have to explain astronomer's observations, which most cosmologists *hope* will turn out to equal zero.")

I have deliberately told this fantastic story before providing the scientific motivation for it, so that my readers may pause to consider just how unbelievable it appears in its first run-through. A subatomic bubble made of false vacuum becomes far larger than anything we can see—before 10^{-30} second has elapsed after the big bang. The false vacuum then re-evolves into true vacuum plus everything else in the universe we encounter today. Can this really be the most popular version of universal history among practicing cosmologists? Are they on to something—or off their beams? To answer these questions, we should examine the arguments in favor of the inflationary theory.

The theory grew out of work by Alan Guth, a particle physicist now at the Massachusetts Institute of Technology, who was seeking to determine what the universe as a whole would be like if he assumed the validity of a class of particle theories called Grand Unified Theories, or GUTs. Since the inflationary model of the universe owes its origins to particle-physics theories invented for independent reasons, it can escape being labeled with the damning words "ad hoc," that is, invented solely for purposes of explaining one particular set of observations.

The primary attraction of the model to cosmologists is that it offers a way to resolve some key unsolved mysteries of the universe—mysteries that have come to be called the *horizon problem* and the *flatness problem*. The horizon and flatness problems may be restated as the pair of questions: Why do regions that have never been in causal contact nevertheless exhibit nearly identical properties? And why is the average density of the universe we observe today so close to the critical density that divides eternal expansion from eventual contraction? That is, why is the universe so nearly flat?

In 1979, when Guth first proposed the inflationary model of the universe, there seemed to be no way to match the model to what we knew—or thought we knew—about the real universe. But other cosmologists, most notably Andrei Linde in Moscow (now at Stanford University) and Paul Steinhardt and Andreas Albrecht of the University of Pennsylvania, soon saw that with a crucial modification, too arcane to explain here, the inflationary theory could serve quite well to explain these two key mysteries of the cosmos.

The Horizon Problem

In the conventional (noninflationary) theory of the expanding universe, there are always parts of the universe too widely separated to be in causal contact, that is, to have exchanged light or any other type of interaction in the time since the big bang. The maximum distance from an observer that light has had time to travel since the big bang equals the age of the universe multiplied by the speed of light. This distance, called the observer's *horizon*, increases in direct proportion to the elapsed time since the big bang. Therefore, as the universe expands (conventionally), we—or any other observer—come into causal contact with an increasing volume of space. In the conventional expanding universe, no observer ever "loses" part of the universe over the horizon; instead, additional volumes of space continuously appear over the horizon of any observer as time increases.

But regions beyond an observer's horizon cannot affect

that observer, because no signals, forces, or impulses have had time to arrive from beyond the horizon. Consider what this means for observations of the cosmic background radiation described in Chapter 6. The cosmic background radiation last interacted with matter at a time about 300,000 years after the big bang. The radiation therefore reaches us from regions almost as distant as our horizon. When we look in two opposite directions, we observe radiation produced in regions that are now separated by *twice* the distance of our horizon. Each of these two regions lies outside the other's horizon, and since they do so now, they must have always done so in the conventional (noninflationary) model of the universe (see Diagram 7). Yet this radiation reaches us from all such opposite directions in nearly the same amount, and with nearly the same spectrum. How is this uniformity possible, if these opposite volumes (as we see them) have never been in causal contact? How do they "know" how to equalize their radiation fluxes, if they have never had the chance to establish contact by exchanging any types of forces or particles?

The answer is that they can't, and don't—if the conventional, noninflationary model is correct. In that case, we must seek a special explanation for the fact that regions of space beyond each other's horizon nonetheless show remarkably similar properties. To date no good special explanation exists. But the inflationary theory offers a direct explanation: The different regions of space all *were* in contact, once upon a time. All the different regions of space within the submicroscopic bubble that inflated did lie within each other's horizon, and did establish a causal interaction, during the tiny amount of time before inflation began. (The mathematically minded will see that since light travels at a speed of 3×10^{10} cm/sec, at a time 10^{-34} second after the big bang, any bubble smaller than about 3×10^{-24} cm— about one ten-billionth the size of a proton—could have had causal contact throughout.) Then, as inflation exploded the bubble into enormous size, the equality of physical properties established during the first 10^{-34} second persisted.

The attentive reader will have noted that this can be so only if inflation causes different regions of the exploding bubble to separate more rapidly than the speed of light, and thus to move outside each other's horizon. And so it is; inflation indeed demands faster-than-light expansion velocities.

"Nothing can go faster than light," one so often hears in life. This aphorism, like so many others, is false. Relativity theory applies the light-speed limit only to "local" velocities, that is, to the speed with which one object can go by another in the same vicinity. In terms of the balloon model of the expanding universe, this means that the balloon can expand without a speed limit; relativity theory merely forbids any two objects to pass one another at any point on the surface with a speed greater than the speed of light, and also forbids the transfer of any information at speeds greater than the speed of light. The inflationary model's successes arise from the requirement, as part of the model, that all parts of the early universe expanded away from one another at speeds far greater than the speed of light.

The Flatness Problem

So much for the horizon problem, so easily solved by the inflationary theory (and by other theories as well, which receive no mention here because they do not deal with the other problems). The flatness problem brings us back to a key parameter describing the universe with which we are already familiar: the average density of matter. We have seen that the exact determination of this parameter presents grave problems, but no one doubts that the *minimum* average density of matter equals the density provided by *visible* matter, the matter that we can see, as contrasted with the total average density, which sums the density of the visible and invisible (dark) matter.

The minimum average density that exists in the universe equals about 2 percent of the critical density, the amount which would mean that space in the universe is flat, and the amount which the universe would require if is ever to cease expanding and start contracting. To ordinary folks living within Earth's biosphere, the factor of 50 by which the density of visible matter falls short of the critical density might seem quite large. But take a moment to consider the fact that the density of matter in the universe *could* be one one-thousandth, one one-millionth, or as little as one one-trillionth of the critical density. Then you may find it remarkable, as many cosmologists do, that the actual density of the matter that we are certain exists comes so close to the critical density. It comes so close, in fact, that to many cosmologists, a special physical explanation seems required; it could not have "just happened" that way, randomly.

The inflationary theory offers such an explanation. Soon after the big bang, inflation made the universe flat. Whatever the curvature of space might have been before inflation, the inordinate expansion during the inflationary era flattened the curvature away. We may analogize inflation to blowing up a balloon from its normal size into something larger than a galaxy. This inflation will smooth away any wrinkles in the balloon, and any modest portion of the balloon will appear very nearly flat. Nearly, but not quite: The inflationary theory does allow for the actual density to differ from the critical density by one part in 10^{40} or so. This difference might ultimately affect the future of the universe, but for now the key result of the inflationary theory, so far as the curvature of space is concerned, is that the theory predicts that the universe must be flat, or nearly so, with a density of matter almost exactly equal to the critical density.

Thus the inflationary model of the universe offers an explanation of the horizon and flatness problems, in a (to many cosmologists) self-coherent, self-consistent way. Inflation also explains why the universe contains no magnetic monopoles—a type of particle which, according to current theories in particle physics, should have been produced in abundance in the early universe but which have never been detected. The theory "inflates" the problem away: It expands space to the point that the number of monopoles per cubic light year falls almost to zero. And as we shall see in Chapter 13, the inflationary universe predicts what seem to be good "seeds" for the formation of structure in the universe. The model therefore does what a model should: explain the hitherto unexplained. And it does more: It predicts definitively that the average density of matter should exactly equal the critical value. Not a bad result! the model fits, explains, and predicts.

But how well does it fit, explain, and predict? Significant objections exist to the model—in addition to the crucial one that evidence for the high density of matter predicted by the inflationary theory is lacking. The recent new results concerning the age of the universe discussed in Chapter 8 make the inflationary theory not quite so appealing as before. In terms of the present value of Hubble's constant, H_0, we have seen that in a universe without a cosmological constant, the time since the big bang has a value between $1/H_0$ and two-thirds of this value. The former time interval will be correct if the universe has an average density much less than the critical value, while the latter value, only 67 percent as large, applies if the density of matter equals the critical density. Thus, in the inflationary model, the universe is only about 8 billion years old—roughly half the age that has been estimated for some of the oldest globular clusters.

If the total density of matter equals 10 to 20 percent of the critical value, as observations now suggest, then the time since the big bang lies relatively close to the value $1/H_0$ that characterizes a low-density universe—at 90 percent of it if the density equals 10 percent of the critical value, and at 80 percent for a density that is 20 percent of the critical value. Thus if $1/H_0$ equals 12 billion years, the time since the big bang equals 10.8 billion years for the 90 percent value and 9.6 billion years for 80 percent. Difficult though these times may be to reconcile with the estimated ages of stars, they nevertheless present less of a problem than the inflationary model, which gives only 8 billion years for the age of the cosmos.

In other words, the inflationary model brings the worst of all worlds, so far as the apparent contradiction between the ages of stars and the time since the big bang is concerned. By itself, this fact does not rule the model out of serious consideration, but we should bear it in mind when inflationary enthusiasts trumpet the virtues of their models. Some inflationary-theory proponents have begun work on variants of the model that do not require that the average density of matter equal the critical density. With this requirement gone, the inflationary model could achieve a much better fit with current estimates of both the average density of matter and the age of the universe. However, these low-density inflationary models specify a nearly exact number of doublings in size during the inflationary period of the cosmos, which seems a bit too "finely tuned" for most cosmologists' tastes. Paul Steinhardt has referred to these attempts to make the inflationary model fit with observational results as "desperate measures for desperate people," which does not, of course, eliminate these measures from the possibility of proving correct.

The best proof of the inflationary model's validity would be the discovery of sufficient dark matter to make it at least likely, or even probable, that the average density of matter in the universe equals the critical density. For this reason, both theorists and experimenters are hard at work searching for dark matter—the theoreticians to invent what it might be, and the experimenters to look for those "might-bes."

Where does the inflationary theory—if it should prove correct—leave our views of the cosmological constant? Inflation requires a nonzero cosmological constant during the early moments after the big bang. The cosmological constant implies that the energy that exists in the false vacuum drives the inflation. But when the false vacuum turns into true vacuum as inflation ends, this cosmological constant disappears.

After inflation, the inflationary model of the universe corresponds exactly to a model without inflation, except that we have solved the flatness and horizon problems and we know that the average density of matter exactly equals the critical density. This does not mean that the cosmological constant must be zero, only that the constant now has the same status that it would if inflation had not occurred. We therefore cannot use the inflationary theory to solve the problems arising from the relatively small amount of time since the big bang in the same way. These problems may be solved by introducing a cosmological constant, but Occam's razor suggests that we adopt such a fudge factor only as a last resort; otherwise we simply replace one mystery with another.

For those who, like Mark Twain, count on science to deliver wholesale returns of conjecture from a trifling investment of fact, the inflationary theory succeeds brilliantly in suggesting unknown realms of speculation. Andrei Linde has emphasized that the inflationary theory implies that "our" universe—everything that we can see or can hope to see— amounts to just one of a much larger (possibly infinite) number of universes contained in a "metauniverse." Each of these universes inflated from a submicroscopic region of space, growing far larger than our visible universe in a time less than 10^{-30} second. The universes did so at random times and in random places, so that one clear implication of the inflationary model is that we cannot hope to say when or where the metauniverse began. Alan Guth has speculated about whether or not you could create a universe in your basement, but this is a trifle fanciful (for the present).

Among the questions that naturally arise in considering the inflating bubbles that each become its own universe, perhaps the most common is, Doesn't the new universe encroach on space in the old one where it appears, posing an onrushing danger to those in its way? The answer turns out to be that no such danger of collision exists, because when a new inflating bubble appears, it creates its own space too. (It will be easier on both the reader and the author if we leave the matter there.) The theory specifies that we can never hope to make contact with the other universes that have come into existence, or are now being born, or will be born during future epochs of our universe, because they are all much farther away than the distance that light has traveled since the big bang that began our own universe. According to inflation, we are captive within our own little bubble, now trillions upon trillions of light years in diameter, and must confine our activities, if not our imagination, to that realm. The "metauniverse" does offer the cheering thought that even though we cannot see them, an uncountable number of other universes exist, each with its own problems, and it certainly implies that the metauniverse has an infinite age and an infinite future extent.

This chapter's discussion of the inflationary model has omitted an especially large number of twists and turns that have on occasion confused even the inflationary experts. Different versions of the inflationary theory have been rejected as too unlikely, even with the loose interpretation of that word that theorists apply, but other versions have then proven more durable. Today, with many versions rejected, different variants of the inflationary model still appear viable. The model variants have in common these properties:

(1) Seemingly empty space, far from being a true vacuum, seethes with energy.
(2) This energy makes a bubble of space expand at truly fantastic speeds during what we call the earliest moments of the cosmos.
(3) As this inflation comes to an end, at a time much far less than one second into the universe's history, the phase transition that produces true vacuum creates enormous numbers of particles and heats the universe tremendously.
(4) After the end of inflation, the universe (once a sub-microscopic bubble of space) expands in just the same way as predicted by the big-bang theory that existed before inflationary models came into being.

Let us take a break from the mind-wrenching demands of the inflationary theory, put aside the contemplation of multiple bubbles in a metauniverse, and turn our attention to a simpler issue: In what form of matter does most of our own universe reside?

ELEVEN

The Mystery of the Missing Mass

Alan Guth's proposal of the inflationary model of the universe burst into cosmologists' consciousness at the end of 1979. Under the careful scrutiny of rival theorists, the inflationary model quickly exhibited some glaring flaws. But almost as rapidly as these emerged, other theorists showed how to modify the inflationary scenario to make everything fit together. One theorist's perfect hypothetical particle became another's fudge factor, but a general consensus emerged that the inflationary model was surprisingly "robust," capable of surviving nearly all theoretical assaults. By the mid-1980s, the inflationary theory had displaced its rivals with startling, though not worldwide, success, so that whenever theoretical cosmologists now turn their attention to explaining the structure of the universe, they usually begin with the inflationary era as a given.

However, many observational astronomers, keenly

aware that data, not theories, describe the real universe, have resisted the inflationary model, which could be compared (with some justice) to medieval cosmologies in which angels danced on pinpoints. Inflationary-minded theorists have been quick to point out that their model not only explains several key aspects of the known universe, such as the flatness and the horizon problems, but also does what a really winning theory should do: It makes a stunning prediction, in this case about the average density of matter in the universe. Go and measure that density, say the theorists, and you will find that it matches the prediction, for the theory appears to be so beautiful that it can hardly be otherwise. But if it is otherwise (they mutter when pressed), then, like good scientists, we must abandon—or modify—our theory.

In throwing down this cosmic gauntlet, the theorists know that a strong observational current has been running in their favor. By the early 1980s all astronomers had come to agree that the universe contains far more invisible matter than the matter that shines in stars. The questions on the minds of all those who cared to probe the depths of the universe were, and remain, does this dark matter amount to 50 times more than the visible matter, thereby fulfilling the inflationary theory's prediction that the average density of the universe equals the critical density? Or does the dark matter provide "only" 5 to 10 times more mass than the visible matter, as observers now estimate it does? In that case dark matter would still dominate the universe, but the inflationary model would seem to be just plain wrong.

The answers to these questions will resolve not only the question of whether the inflationary theory is correct about the beginning of the universe but also whether eternal expansion or eventual contraction lies in store in the future. For this reason, we must take an excursion through the evidence, accumulated during the 1960s and 1970s, that compels even the most empirically minded observational astronomer to conclude that dark matter dominates the universe.

Searching for the Invisible

When we look at the points of light that spangle the night skies, we see matter of a particular type, arranged in particular ways. Except for the nearest and smallest of these points of light—our moon and the other planets that orbit the sun—nearly every object that shines is either a star, a mass of gas that will become a star cluster, an actual cluster of stars, a galaxy, or matter expelled from stars. The apparent exceptions to this rule include two categories of objects, the quasars, which are probably galaxies in the process of formation, and the gamma-ray bursters (recently detected in the thousands by new satellites above our atmosphere), which emit streams of high-energy photons through processes now undetermined. In our own galaxy, we must also include interstellar matter, which floats among the stars, occasionally giving birth to groups of new stars. Since interstellar matter emits radio waves, it cannot qualify as dark matter, and we know that in a giant spiral galaxy such as the Milky Way, the total mass of the interstellar matter amounts to only a few percent of the mass contained in stars.

By definition, dark matter emits neither visible light nor any other form of electromagnetic radiation that we can detect. If astronomers can't "see" it, how can they study it?

Indeed, how do they even know it's there? The answer, as with so many things in physics, is gravity. The dark matter has revealed itself through the operation of gravity, the force that rules the cosmos at the largest scales of distance. To understand astronomers' detection of enormous quantities of invisible matter, we must therefore come to understand, and thus to appreciate, the properties of gravitational forces.

Modern physicists rank gravity among four basic types of forces. The other three, as we have seen, are the strong force and the weak force (both of which govern the interactions of elementary particles when they approach each other to within extremely small distances) and electromagnetism (which rules the ways that charged particles affect one another at distances too great for strong weak forces to apply). Two of the forces—gravity and electromagnetism—act over large distances, whereas the other two, the strong and weak forces, have essentially zero effect between any particles separated by a distance larger than the size of an elementary particle (approximately 10^{-13} centimeter).

For that reason, when scientists look at the big picture, gravity and electromagnetism naturally draw attention. Both types of forces behave the same way with increasing distance: The strength of the force between any two objects decreases in proportion to the square of the distance between them. For gravity, the amount of force between two objects varies with the product of the *masses* of the two objects. Electromagnetic forces, in contrast, care nothing for objects' masses; instead, the amount of electromagnetic force between two objects increases in proportion to the product of the objects' *electric charges*.

This difference allows gravity to dominate the universe at large distance scales. Electric charges come in two varieties, positive and negative, and the electromagnetic force between two objects can likewise be either attractive (between a positively and a negatively charged object) or repulsive (between two positively charged or two negatively charged objects). But mass comes in only one "charge," so that gravitational forces always attract and never repel. (Negative mass, or at least "negative gravity," plays a role in more than one novel, but physicists do not regard the concept as a serious possibility for the real world. Negative gravity should not, of course, be confused with false vacuum.)

As a result, when we encounter progressively more massive objects, we meet objects that produce greater amounts of gravitational force than less massive objects do at the same distance. These objects, however, are not likely to exert larger electromagnetic forces than smaller ones, for the excellent reason that they tend to be made of nearly equal numbers of positive and negative electric charges, and therefore produce essentially *zero* amounts of electromagnetic force.

What produces this equality of charge? If an object happens to acquire a preponderance of one type of charge—positive, for instance—it will attract negative charges, and repel positive ones, through electromagnetic forces. Eventually, the object will collect sufficient additional negative charge to become electrically neutral, simply because electric charges, and electromagnetic forces, come in two varieties.

So gravity governs the interactions of cosmic objects;

it dominates their motions and exerts a powerful influence on their evolution as individuals. In a star like our sun, all four types of forces are seriously engaged. Gravity holds the star together, despite the fact that what amounts to a trillion hydrogen bombs are exploding each second in its interior. At the high temperatures within the star, atoms are stripped into negatively charged electrons and positively charged nuclei, despite the electromagnetic forces that keep the electrons orbiting around the nuclei in cooler regions. The fast-moving nuclei tend to repel one another through electromagnetic forces, and only in the star's hottest part, its central core, do they manage to approach one another closely enough for strong and weak forces to make the nuclei fuse together. This fusion turns some of the mass in the nuclei into energy. More precisely, fusion converts energy of mass—the energy locked within every particle with mass, in an amount equal to the mass times the square of the speed of light—into kinetic energy (energy of motion). The kinetic energy diffuses outward through countless collisions, heating the entire star and making its surface glow with the starlight that we see.

But when we consider the forces that govern how a star interacts with its surroundings, gravity alone merits close attention. Gravity keeps the sun's planets in orbit, while electromagnetic, strong, and weak forces from the sun amount to nothing whatsoever. Similarly, when we ask how the sun affects the movement of other stars in the Milky Way galaxy, or how those other stars affect the movement of the sun, the answer lies in gravity.

Secrets of the Motions of Stars

Once astronomers realized that our solar system belongs to a spiral galaxy, they devoted much effort to mapping the details of this giant spiral and to investigating how the stars move within the Milky Way. This was not an easy task: Living on a planet moving in orbit around an individual star, we must attempt to map the motions of other stars while simultaneously determining—and allowing for—the dual motions of our own star and of our planet around that star. Despite the difficulties of these measurements, a startling result slowly became evident. The movement of stars within the Milky Way suggests that the galaxy contains far more mass than can be accounted for by the 300 billion or so stars that the galaxy contains.

From where does this conclusion arise? From a knowledge of gravity, plus careful measurements of the speeds at which stars are moving. As we discussed earlier, astronomers employ the Doppler effect to determine the relative motion of an object toward us or away from us. Greater amounts of motion produce larger Doppler shifts in the wavelengths and frequencies that we measure on Earth in the spectrum of light emitted by the object. If we know what the object's spectrum would be if no motion occurred, these Doppler shifts reveal the object's relative velocity along our line of sight. Astronomers have come to recognize a host of particular patterns, each associated with a particular type of star, in the spectrum of absorption and emission lines. They can recognize these patterns even when all the lines have been shifted by the Doppler effect. They can therefore compare each star's observed spectrum with the same, unshifted, spectral pattern, and calculate the amount of the Doppler shift.

Starting early in this century, astronomers began to assemble Doppler-shift data for the stars in the Milky Way. Statistical analysis of these data slowly revealed that most stars, including our sun, move in nearly circular orbits around the galactic center. Astronomers knew that since gravity dominates the forces in the Milky Way, each star's motion arises from its response to the gravitational forces from *all* the other stars in our galaxy. At first thought, it might seem a herculean, if not an impossible, task to calculate the forces from 300 billion other stars. However, a sweet trick, first demonstrated by Isaac Newton, makes this calculation easy. (Newton was dealing not with the Milky Way but rather with the gravitational force inside a spherical distribution of matter. His trick, however, works well in both situations.)

The trick is this. Consider any star—for example, our own sun, located about 30,000 light years from the center of the Milky Way. Mentally divide all the stars in the galaxy into two groups: those closer to the galactic center than the sun, and those more distant from the center. Assume, as seems reasonable, that the distribution of stars in the Milky Way favors no one direction outward from the center, so that the numbers of stars in each direction are approximately equal. Then, Newton showed, each of the two groups produces a simple effect on the sun.

The gravitational forces from all the stars closer than the sun to the galactic center combine to pull the sun directly toward the center. This pull exactly equals the force from a single object at the center whose mass equals the sum of the masses of all these stars. Hence we can (mentally) replace all these billions of stars with an imaginary object embracing all the mass closer to the galactic center, located exactly at the center.

And what of the myriad gravitational forces from the stars farther from the center? Newton showed that an even greater simplification rules here, so long as the stars have a spherically symmetric distribution around the galactic center. The gravitational forces from the stars farther from the center cancel one another, producing a zero net force on the sun.

Newton's analysis does not depend on the type of objects that exert gravitational forces, and remains valid despite Einstein's refinement of Newton's theory of gravitation. So long as the objects favor no particular direction in their distribution, *all* objects of whatever type that lie closer to the center than the sun combine their forces as described above, and all objects farther from the center combine to exert no net force on the sun. Because we have used the sun as a representative star, the same statements will be true for any star in the galaxy, so long as we divide everything else into objects lying closer to the center and farther from the center than *that* star. Newton's simplification applies perfectly only to a spherical distribution of matter, and the Milky Way is far from being spherical. Nevertheless, because the Milky Way possesses a symmetry around the axis that the stars orbit in circles, the simplification works fairly well, and astronomers know how to adjust for the details and to determine the gravitational force that the totality of mass of our galaxy exerts on the sun.

Newton's insight has been used to advantage for the last five decades by astronomers who seek to measure the amount and distribution of mass within the Milky Way. The basic observational strategy has been to use the Doppler effect to measure the motions of stars and glowing gas clouds at different distances from the galactic center. Thanks to Newton, we know that the speed in orbit around the center of any object depends on two quantities: its distance from the center, and the amount of mass that lies closer to the center. Newton's law of gravitation provides the amount of force on the object, and Newton's second law of motion tells us how the object will react to this force. Since we observe the object's reaction—its speed in orbit—and since we know its distance from the galactic center, we can determine the missing parameter, the amount of mass that lies closer to the center.

When astronomers observed the motions of objects closer to the center than our sun, they obtained results that surprised no one. The amount of mass closer to the center than the stars and gas clouds turned out about equal to the mass that we would obtain by estimating the number of stars closer to the center and multiplying by the average mass per star (about 70 percent of the sun's mass). Thus, up to our distance from the center, the mass in the Milky Way resides mainly in stars, with a smaller contribution from interstellar clouds of gas, as everyone had expected.

But as astronomers turned their attention to regions of the Milky Way significantly more distant from the center than our sun's 30,000 light years, and measured the speeds of stars in the disk, their results were startling. The astronomers found that the speeds of these more distant stars do not decrease, implying that the parts of the Milky Way between those stars and our sun contain far more mass than can be found in the stars that we see. Although relatively poor in stars, our galaxy's outer "halo"—a spherical distribution surrounding the basic plane of the Milky Way but many times larger—has a total mass far greater than the mass of all the visible stars in the Milky Way! (See Diagram 12.)

This conclusion is strengthened by mathematical analysis of what happens to a flattened, rotating disk of stars over billions of years, as stars circle the galactic center dozens of times. Unless the galaxy has far more mass *beyond* the disk than the disk itself contains, after only a few rotations the disk will warp and twist, no longer presenting an orderly, flattened appearance. But a massive halo, much larger than the disk, will stabilize the situation, allowing the disk to maintain itself for billions of years.

As the observations and calculations described above received steady confirmation among astronomical experts, the message grew abundantly clear. The invisible hand of gravity, working on the motions of stars in the Milky Way, has revealed enormous amounts of invisible matter that constitute the dominant mass of our galaxy. Dark matter (it is worth repeating) emits not only no light but also no radiation at any frequency of the electromagnetic spectrum that astronomers study—and by now, with the help of telescopes orbiting above the atmosphere, they can study nearly all of them.

When the notion of dark matter first appeared, it was used to explain not the motions of stars in the Milky Way but movement on a much larger scale: the motions of individual galaxies in galaxy clusters. Sixty years ago, the Swiss-Bulgarian astronomer Fritz Zwicky, who had become a professor at the California Institute of Technology (and who was one of the Americans delegated to interrogate the German physicists who had worked for the Nazis on their atomic bomb project), promoted the idea that much of the mass in galaxy clusters consists of matter that does not shine.

Zwicky derived his ideas from Doppler-shift measurements of the motions of the galaxies that form the Coma cluster described in Chapter 7. He estimated the average separation of the galaxies, and used the Doppler effect to measure their motions toward us or away from us. The average separation and the line-of-sight velocities allowed him to deduce the total mass in the Coma cluster. The velocities turned out to be significantly larger than would have been estimated from the masses of the galaxies alone: The total mass was *several hundred times more* than the mass we would assign to the cluster by adding together the estimated mass of each galaxy, deduced from the brightness of its stars and our knowledge of the mass that stars contain.

During the 1950s and early 1960s Zwicky's suggestions were almost completely ignored, the result as much of his style of argument as of the perceived flaws in his ideas. Were Zwicky alive today, he would be more than delighted to underline his prescience, for his concept of unseen matter has been taken quite seriously by astronomers during the past 25 years—just about the time, by coincidence, that

Dark Matter throughout the Universe

Zwicky has no longer been in equilibrium with the biosphere. Evidence completely unavailable in Zwicky's time has now tipped the balance in favor of proving dark matter's existence.

What additional new evidence suggests the existence of dark matter on larger scales of distance than our own Milky Way? First of all, we may reason from the "averageness" of the Milky Way galaxy—though strictly speaking, this can hardly count as true evidence. The Milky Way ranks, in astronomers' eyes, as an almost prototypical giant spiral galaxy. In its size, its number of stars, its overall structure, and the general pattern of motions of its stars, the Milky Way closely resembles what we can determine about other giant spirals. What holds true for the Milky Way should therefore be true of its cousins as well.

Furthermore, examination of nearby galaxies has confirmed this mental extrapolation. For the past 25 years, many astronomers have analyzed the motions of the outermost parts of nearby giant spiral galaxies. This is not an easy job, because the sparseness of stars in these outer regions makes it difficult to obtain sufficient light for spectroscopic measurements. Radio measurements of the motions of gas clouds have supplemented the optical measurements of stellar velocities. The results prove similar to those found for the Milky Way. These galaxies also seem to possess enormous, massive halos of dark matter. Similar conclusions, though less well established, have been drawn from studies of the motions of objects and the emission of radio waves and x rays in the other type of giant galaxy, the ellipticals.

And if additional confirmation were needed, that too

has appeared from studies of relatively rare situations in which two galaxies close to each other orbit around their common center of mass. By measuring the galaxies' speeds in orbit (using the Doppler effect) and the distance between them, astronomers can calculate the masses of the moving galaxies. Those masses also turn out to be many times greater than the mass that we see shining in the galaxies' stars. One example of this appears close to home: The Milky Way and the Andromeda galaxy, by far the two largest galaxies in the Local Group, are approaching each other with a relative velocity of 270 kilometers per second. This high velocity provides additional evidence that these two galaxies have far more mass than can be accounted for by their stars.

Of course, exceptions exist to nearly every generalization about our observations of the universe. In 1993 astronomers reported that in the relatively nearby giant elliptical galaxy called M105, objects' motions reflect an amount of mass that can be explained by the stars that we see! The irony is that the dark matter has become so well established that the foregoing sentence made (astronomical) news. For once, astronomers had found a galaxy in which the velocities of objects far from the center decrease in just the manner expected if the stars contain the bulk of the galaxy's mass.

As Fritz Zwicky first demonstrated, the technique of determining the total mass of an object from the motions of its component parts applies not only to individual galaxies but also to clusters of galaxies. Here too gravity provides the invisible glue that binds the cluster's individual member galaxies, which move at speeds that increase as the total mass in the cluster increases. Just as in the case of giant spiral galaxies, evidence has slowly accumulated during the past decades to show that most "rich" clusters contain much more mass than can be explained by the masses of the stars in the individual galaxies.

During the past three decades, a host of astronomers have used the world's largest telescopes to observe cluster after cluster, improving and expanding upon the original work by Zwicky. For most of these galaxy clusters, the total mass seems to be about 5 to 10 times greater than can be explained by the starlight we see in their thousands of member galaxies. Thus, in large clusters of galaxies, as in individual giant galaxies, measurements of objects' motions imply that dark matter dominates visible matter by a factor of 5 to 10, more or less. This confirms Zwicky's original estimate, made from a much more limited set of data.

The effects of gravity in galaxy clusters can now be discerned in more ways than the study of objects' motions by means of the Doppler effect. In 1992 an x-ray observing satellite called ROSAT (a joint project of Germany, the United States, and the United Kingdom, named the Roentgen Satellite after Wilhelm Roentgen, the discoverer of x rays) turned its detectors toward the constellation Cepheus, to study a group of three galaxies named the NGC 2300 group after the most prominent member galaxy. ROSAT revealed that this group of galaxies lies immersed within an x-ray emitting region more than a million light years across. This region emits about ten billion times more energy each second in the form of x rays than the sun produces per second in the energy of its visible-light output.

Observations of this x-ray emitting material, which is

certainly *not* dark matter, allow astronomers to deduce the
amount of dark matter that permeates the group of galaxies.
At least five steps exist in the chain of deduction that links
the observation of x rays to the fact that dark matter dominates even this small group of galaxies. First, x rays are the
sign of hot gas. Second, the amount of x-ray emission reveals
how much hot gas exists, and how high its temperature is.
Third, since hot gas tends to escape, we can calculate the
amount of mass that the cluster must contain in order to
keep the hot gas from evaporating. Fourth, we can estimate
the total mass contained in the stars that make the galaxies
shine. Fifth, we can compare this visible mass with the
amount of mass deduced to exist in step three, and thus
discover that the latter amount is much larger—so that most
of the matter in the group of galaxies must be dark matter.

When this series of steps is applied to the NGC 2300
group, we find that the x-ray emitting gas has a total mass
of 600 billion solar masses and a temperature of about 15
million degrees Fahrenheit. To retain this much gas at such
a high temperature, the group of galaxies must have a total
mass equal to 30 trillion times the mass of the sun. If we
assume that the average mass per star has the same value in
the three galaxies in the NGC 2300 group that it does in the
Milky Way (70 percent of the sun's mass), the total mass in
the stars of all three galaxies equals about 600 billion times
the mass of the sun. Thus, the NGC 2300 group apparently
has about 50 times more mass in dark matter than it does
in stars.

In interpreting a result such as this, astronomers are
well aware that they must include allowances for possible

errors. For example, they do not know the distances to the galaxies in the NGC 2300 group exactly. An error in the distance will lead to an error in the estimated amount of hot gas, and hence to an error in the total amount of mass needed to prevent the gas from escaping. In addition, they may have made errors in determining the amount of x-ray emission. These errors could reduce the deduced ratio of the amount of dark matter to the amount of visible matter by a factor of 5 or more. But even a ratio of 10 to 1 in the masses of dark matter to visible matter would testify to the domination of the group of galaxies by dark matter, and in approximately the same ratio as found for giant spiral galaxies and rich clusters of galaxies.

As we discussed in Chapter 9, we can determine the total density of matter over distances much greater than those spanned by galaxy clusters by finding large-scale flows that deviate from the general Hubble flow, the universal expansion. A recent analysis by Marc Davis—the astronomer who began the first redshift survey at Harvard and is now a professor at the University of California, Berkeley—concludes that these flows do exist. Davis's models of the flows require that the average density of matter have a value that implies at least 10 times more dark matter than visible matter. Without this density, deviations in the amount of matter from region to region would not lead to the statistically determined amounts of flow. Unfortunately, Davis cannot yet determine whether the average density of matter lies closer to the critical density than to values that are "only" 10 to 30 percent of the critical density.

This difference has crucial cosmological significance.

First of all, if improved models and observations show that the density equals "only" 10 to 30 percent of the critical density, then the inflationary model falls onto evil times; on the other hand, a density close to the critical density validates the inflationary theory. Second, the result provides the strongest evidence today that dark matter dominates the universe, providing at least 10 times more of the total density than familiar types of matter, over the largest distances scales accessible to observation. Striking though the results from galaxy clusters may be, it might have turned out that the dark matter dominated only relatively small regions of space. Davis's study buttresses the notion that dark matter spreads through the entire universe, dominating familiar matter everywhere by a factor of 10 or so.

In addition to all of the observational evidence astronomers have amassed for the existence of dark matter, the continuing discoveries of particle physics have made theoreticians more receptive to the notion that we have yet to discover the particles that form most of the universe. To physicists who specialize in theories of different types of elementary particles, the existence of the dark matter has been a clarion call from the darkness, and a spur to their inventiveness.

We should not lose sight of the fact that although these theorists must attempt to tell us what type of matter provides most of the universe, this does not mean that they have to come up with enough dark matter for the average density to equal the critical density. The point deserves emphasis because of theorists' enthusiasm: Most of them find the inflationary theory so appealing that they have no plans to truncate their hypothetical dark matter at a density of "only" 10 or 20 or 30 percent of the critical density. Cooler heads are needed to judge just how much dark matter can be furnished by a particular hypothetical component of the universe. Those cooler heads belong to experimental physicists, who may someday find evidence for a particular type of particle, or disprove that the particle type exists in sufficient numbers to provide the dark matter. For now, with those experimental tests only beginning, most opportunities for hypothetical dark-matter particles to comprise most of the cosmos remain wide open—so long as the particles are not ordinary matter.

So the impending millennium finds astronomers in the peculiar situation of having recently discovered most of the universe, without knowing what it is. Dark matter is *the* cosmic mystery, which may be cleared up during the next few years or may outlive us all. But precisely because dark matter does exist—almost no one disputes that—and because it dominates ordinary matter in mass and thus in gravitational effect, astronomers will seek to find it; theoretical cosmologists will continue to use it in their models; and particle physicists will go on speculating about the forms it may take and the methods by which it may be detected. We can take a chapter to see how widely speculation has ranged—and another to see how useful it has proven to be.

TWELVE

In Search of Most of the Universe

Today astronomers accept the existence of dark matter as the dominant part of the universe, furnishing at least 5 to 10 times more matter, and possibly 50 times more, than the familiar matter that we see shining in stars, forming new stars, collecting into meteoroidal dust, or accreting to form planets like our Earth. But acceptance of this overwhelming predominance of dark matter did not occur overnight; astronomers and cosmologists rightly resisted such a large alteration to our concept of the universe without strong evidence. As the evidence grew stronger, however, astronomers gradually shifted into a pro-dark-matter view. Quite naturally, they began to ask the question, What and where is the matter?

As to the issue of where the dark matter resides, the answer seems clear: everywhere. The luminous matter we see seems to be a light frosting on the basic dark cosmic cake.

Astronomers feel sure from cosmological theories that in the early universe all types of matter were distributed almost evenly throughout all of space. Since then, the effects that caused matter to form clumps may have acted more strongly on the visible matter than on the dark matter, since we find the dark matter to be spread out more homogeneously than the matter that shines, as in our own Milky Way, for instance. Both dark and visible matter are clumped, but the matter we can see is more clumped than the matter we can't see.

So if the dark matter is everywhere, what is it? An embarrassment of riches floats on the river of answers to that question. From the minds of theoretical physicists have come a host of possible explanations for the dark matter, including (in a truncated listing) hypothetical particles called axions, cosmic strings, domain walls, gluinos, gravitinos, higgsinos, magnetic monopoles, majorons, massive neutrinos, maximons, neutrilinos, paraphotons, photinos, preons, quark nuggets, shadow matter, sneutrinos, wimps, and winos. Each of these particle types differs from the others in such properties as the particle's mass and its ability to interact with other types of particles. Each of these types might exist; no theorist (especially those promoting a particular type of particle) dares to allow the possibility that all of them exist, leaving Occam's razor a lump of twisted metal in the cosmic wasteheap.

In plain fact, astronomers and cosmologists have no good explanation of what the dark matter may be. Actually, they have dozens of "good" explanations—they only lack agreement on which, if any, is the correct one. What can be done to determine what most of the universe is made of?

Science has a procedure that has proven immensely useful: classify, and seek thereby to simplify. By attempting to sort nature's bounty (or even the output from theoreticians of the cosmos) into different piles, scientists hope to see how to separate the wheat from the chaff—and how to make sure they can tell which is which.

To classify candidates for the bulk of the dark matter, we must identify the properties of different types of particles that have the greatest impact on the universe that we can examine. Other particle properties are less significant, since they produce lesser effects on "our" universe, the universe accessible to our observation. And what are these most important properties? According to the best thinking of modern cosmologists, they are twofold: baryonic versus non-baryonic, and hot dark matter versus cold dark matter.

Baryonic or Nonbaryonic?

Before we can appreciate the crucial importance of baryonic versus nonbaryonic matter, we must pause to define our terms. The word "baryonic" (pronounced to rhyme with "scary-onic") does not trip lightly on most tongues, yet it has much to tell us about the universe. Indeed, probably not one in a dozen physicists could describe the etymology of this word—nor explain how this etymology might confuse us as to the word's current meaning among cosmologists.

Baryonic is the adjectival form of baryon, meaning a "heavy particle," whose root is the Greek word *barus*, meaning heavy. The same root appears in the word baritone (and, in a less recognizable form, in the words brigade and grave). Physicists originally called certain particles baryons in order to distinguish them from lighter particles, which they named *leptons*, after the Greek word *leptos*, meaning fine or small. (A lepton is also a Greek coin worth only 1/100 of a drachma—too nearly worthless to mint nowadays.)

Electrons and neutrinos are good examples of leptons; representative baryons are the protons and neutrons that form the nuclei of all atoms. Aside from the particle masses (low for leptons, high for baryons), the most significant difference between baryons and leptons appears when we consider particle interactions: Baryons participate in strong interactions; that is, they "feel" the strong force. Leptons, in contrast, are immune to the strong force, and simply do not experience any effect from it. Hence, the strong force can bind together protons and neutrons to form atomic nuclei, but can never make electrons or other leptons join those nuclei.

Given this history, one might expect that physicists would use the phrase "baryonic matter" to signify protons and neutrons taken collectively, in contrast with "leptonic matter," which would stand for electrons, neutrinos, muons, and other collections of leptons. Such a conclusion would seriously misstate the actual usage. Through the past few decades of cosmological intercourse, baryonic matter has come to mean not heavy particles as opposed to lighter ones but rather all familiar particles—baryons *and* leptons—as opposed to all hypothesized, not yet discovered forms of matter. Baryonic matter participates in the interactions governed by the strong force, but the hypothesized nonbaryonic matter does not.

Something seems wrong here. Didn't we just say that leptons don't feel the strong force? Why aren't they nonbaryonic matter? Humpty Dumpty provides the answer: The words mean what physicists say they mean, no more, no less. Baryons—protons and neutrons, along with more exotic types of heavy particles—dominate the world of particle physics, precisely because they interact through the strong force. These interactions provide the grist for the mills (some costing a few billion dollars) that dominate the activities and the dreams of experimental high-energy physicists. So when these physicists sought a name for their hypothesized new forms of matter, to them it seemed natural to use the phrase "nonbaryonic" to distinguish what might otherwise have been named "unfamiliar" or "truly weird" matter from more familiar types.

Physicists had already assigned terms such as "strange," "charmed," and "flavored" to different subsets of elementary particles; perhaps they sensed the advantages of modesty. In any event, they have laconically named the strangest, most unfamiliar, most purely hypothetical (so far) types of matter "nonbaryonic." If it's baryonic matter, it includes both baryons and the leptons that are associated with them, so the word "baryonic" describes all of the particles that physicists have yet discovered (with the exception of neutrinos and antineutrinos, discussed below). Nonbaryonic matter, however, remains to be found, but whatever it may be (including the possibility of remaining forever a theoretical concept only), nonbaryonic matter never interacts with anything through strong forces.

The Density of Baryonic Matter in the Universe

That nonbaryonic matter cannot feel strong forces is a crucial point, because it allows us to rule out of contention for the dark-matter prize all protons, electrons, atoms, molecules, dust, rocks, asteroids, planets, and everything else that we (rightly) think of as dark, simply because it does not shine. From its definition, we know that baryonic matter would have participated in the great era of *nucleogenesis,* the first few minutes of furious nuclear fusion after the big bang, when the universe had temperatures of a billion degrees or more. During that long-vanished era, the particles that comprise ordinary (baryonic) matter collided with such fury that they often fused together through the effect of strong forces.

The era of nucleogenesis left behind a cosmic fossil that can reveal a key piece of information: the relative abundances of the simplest nuclei in the universe. Calculations show that the first few minutes of the universe left most of the mass of the baryonic matter in the form of individual protons. But about 10 percent of the baryonic matter, counting by mass, fused to form helium 4 nuclei, with two protons and two neutrons. Much much smaller amounts of baryonic matter were left in the form of two other nuclear isotopes of hydrogen, deuterium (with one proton and one neutron) and tritium (one proton and two neutrons), and one isotope of helium, ^{3}He (with two protons and one neutron). The abundances of deuterium and ^{3}He nuclei, along with measurements of the abundances of the two isotopes of lithium, ^{6}Li and ^{7}Li, can provide astronomers with the key to determining the density of baryonic matter in the universe. Fantastic though it may appear, the science of nucleogenesis has developed to the point that we understand almost exactly how much of these isotopes must have been made during the universe's first few minutes, provided that we know one key parameter.

That parameter is the average density of baryonic matter during the era of nucleogenesis, which determines how much of each isotope emerged from the first few minutes. Not only could the early universe form nuclei; it could also fuse them into other nuclear species. Hence the small amounts of deuterium, ^{3}He, ^{6}Li, and ^{7}Li left behind from the first few minutes represent the net result of the cosmic fusion that formed and destroyed these nuclei throughout the universe. Calculations show that the greater the density of matter during the first few minutes, the smaller would be the amount of deuterium, ^{3}He, ^{6}Li, and ^{7}Li left behind (see Diagram 10). Higher densities would have increased the efficiency with which the universe fused these nuclei into heavier ones. Thus the calculated abundances of deuterium, ^{3}He, ^{6}Li, and ^{7}Li turn out to be highly sensitive to the density of matter during the first few minutes. In contrast, the abundances of ^{1}H (protons) and ^{4}He nuclei are relatively insensitive to the density of matter in the early universe. Since the average density of matter in the universe now can be directly related to the density at any moment during those first few minutes, astronomers have long realized that measuring the abundance of these light nuclei offers an excellent path to determining the density of baryonic matter—a number that could determine the future of the universe.

But in order to make this method work, astronomers require a sample of the universe that has remained far from any stars. Stars "cook" deuterium, ^{3}He, and lithium into other types of nuclei, much as the early universe did, and thereby fry our attempts to find out how much of these nuclei the early universe made. We should not despise stars for this activity, because nuclear fusion within stars has produced the vast panoply of nuclei heavier than helium. Had this not occurred, we would not be living on a planet made of these heavier nuclei. However, nuclear fusion within stars implies that we would be unwise to measure the abundance of nuclei in the Earth or the sun and then assume that we have found the amounts made during the era of nucleogenesis. Astronomers have used this method, but only by assuming the absence of "star pollution" can they draw tentative conclusions about the density of baryonic matter in the universe.

Ideally, astronomers would like to measure the abundances of deuterium (^{2}H) and ^{3}He nuclei in intergalactic space, in material that has never condensed into stars or even been close to them. The best modern telescopes, such as the 10-meter Keck Telescope on Mauna Kea in Hawaii, have been used for such observations, but no conclusive results have yet emerged. Within the next few years, however, measurements of intergalactic ^{2}H and ^{3}He should become available, and should provide a firm determination of the density of baryonic matter.

If we rely on the currently available measurements of abundances of different types of nuclei, and assume that astronomers have found places where these abundances accurately reflect those that emerged from the first hours after the big bang, we may draw a firm conclusion about the density of baryonic matter. To the extent that we have made a representative sample of the universe, baryonic matter can provide no more than 4 to 8 percent (probably closer to the lower figure) of the critical density needed to close the universe (see Diagram 10). Even though the "ifs" listed above must be borne in mind, this result accurately states the majority conclusion among cosmologists. Therefore, when we encounter any density of matter much greater than 2 percent of the critical density, we may feel reasonably sure that we are talking about nonbaryonic matter, not about the stuff of which stars, planets, or people are made.

Despite astronomers' reluctance to accept that most of the universe consists of matter completely unlike anything we know, the requirement that the cosmos must consist mainly of nonbaryonic matter agrees with the observations made in 1994 by the Hubble Space Telescope, which almost completely ruled out the most straightforward baryonic possibility for the dark matter, the faint red dwarf stars. The next most popular baryonic candidates are smaller dark objects, such as rocks, comets, or planets. Recent observations of "machos" (massive compact halo objects, described in Chapter 14) tend to rule them out as well as the main component of the dark matter; in any case, since astronomers detect machos only by the effects of their gravitational forces, they might be nonbaryonic rather than baryonic matter.

If further observations verify present theories about the abundance of deuterium, the bottom line will be this: Any hopes for reversing the universal expansion must reside with the dark matter, and this dark matter must be nonbaryonic, that is, made of matter utterly unlike anything we have studied until now in the cosmos. Diagram 11 shows the restrictions that our current observations place on the possible values of H_0 and the ratio of the average density of matter to the critical density.

So, What Is the Dark Matter?

Having established what the dark matter is not—baryonic matter—we run up against the question of what it is. Here the long list on page 164 comes into play. If we employ once again our plan of classification, a single fact about all the dark-matter candidates leaps into prominence: One of them is known to exist, while the others are waiting for confirmation as anything more than theorists' pet notions.

The particle that exists is the *neutrino,* along with its similar but slightly different antiparticle, the antineutrino. In actual fact, physicists recognize three types of neutrinos and three associated antineutrinos, but for our purposes we can consider them all as a single type of particle. Neutrinos and antineutrinos are leptons. They experience neither the strong force nor (since they have no electric charge) the electromagnetic force. They do feel the effects of the weak force, but neutrinos demonstrate just how weak this force is. Neutrinos generated by the trillions upon trillions every second in the center of the sun pass directly outward through nearly a million kilometers of overlying matter, with only the tiniest fraction ever undergoing any interaction with one of the particles in the sun. One one-billionth of these solar-generated neutrinos hits Earth and passes through it at the speed of light (and through us, if we are in the way). To capture even a few of them takes hard-working physicists a great deal of effort.

Neutrinos were predicted to exist long before they were actually detected. In 1932 Wolfgang Pauli proposed the existence of a hitherto undiscovered particle to explain otherwise mysterious experimental results. Pauli's hypothetical particle, named the neutrino (little neutral one) by the Italian physicist Enrico Fermi, received final observational confirmation in 1956 in particle-accelerator experiments. The fact that the existence of neutrinos has been repeatedly verified gives them a unique position among all the dark-matter candidates. (Of course, theorists who propose new and exotic particles take inspiration from Pauli, who had to wait 24 years to see his prediction confirmed by direct evidence.) If neutrinos have zero mass, they cannot have much to do with the dark matter. But if each neutrino has even a tiny nonzero mass, the situation could change dramatically.

Early in 1995, physicists at the Los Alamos National Laboratory in New Mexico announced that they had found evidence for a neutrino mass somewhere between one one-millionth and one one-hundred-thousandth of the mass of an electron. Other physicists were properly skeptical, since the report first appeared in the *New York Times* and not in the physics journals—a violation of protocol that sets back the ears of many scientists. All agreed that the reported result would be highly significant if true, especially if the neutrino mass were confirmed at the high end of the mass range announced from Los Alamos.

Since the universe contains about a billion times more neutrinos than electrons or protons, and since an electron mass equals 1/1,836 of a proton mass, if each neutrino has a mass equal to 1/100,000 of an electron mass, then the total mass of the neutrinos roughly equals 5 times the total mass of the protons, which form the (hitherto) dominant mass component of baryonic matter. Since the mass of helium nuclei equals about one-third of the mass in protons, and since protons and helium together form nearly all the mass of ordinary matter, a neutrino mass at the high end of the announced range implies that neutrinos possess about 4 times more mass than all forms of ordinary matter familiar to us from stars and galaxies.

In this case, neutrinos might represent most of the dark matter—provided that the total density of matter equals only about 10 percent of the critical density. Stated another way, neutrinos cannot give the universe sufficient density to satisfy the inflationary model, or to make the universe eventually contract, even if their mass turns out to equal fully 1/100,000 of an electron mass. But neutrinos could be the dark matter, if it turns out that the density of dark matter is only about 4 times greater than the density of visible matter. However, as we discussed on page 158, the density of dark matter now seems to be at least about 10 times greater than the density of visible matter. In that case, neutrinos might furnish as much as about half of the dark matter, but no more. This would be advantageous for the formation of galaxies, since, as we discuss in Chapter 14, astronomers require other types of dark matter than neutrinos to model the formation process successfully.

What an odd situation! The one particle we know to exist in enormous numbers might turn out to be half the dark matter—but only half! The astute reader will have long since grasped the basic principle of cosmology-watchers:

Wait for the confirmation of experimental results, and don't hurry to believe that everything a cosmologist says is true. If the mass of neutrinos proves insufficient to explain the dark matter, cosmologists will continue to consider hypothetical candidates for the leading role of dark-matter particle. These candidates share the label of not-yet-and-perhaps-never-to-be-discovered. Each type may be the apple of a theorist's eye, but these dark-matter candidates may nevertheless be doomed to a merely theoretical existence in some model universe that will turn out not to correspond to reality.

All of these hypothesized but not yet detected particles share the common property of nonbaryonic matter: no interactions through the strong forces. What possibilities for interactions that might allow us to detect them does that leave? Only the electromagnetic force and the gravitational force—at least among the types of forces that we know. The electromagnetic force can also be ruled out, because modern physics has shown that the weak and electromagnetic forces are so intimately interrelated that if particles can't interact through the weak force, they can't interact through the electromagnetic force either.

That leaves gravity—and a good thing, too. If nonbaryonic matter did not experience the gravitational force, we would hardly have wasted our time carefully defining it. The great hope for nonbaryonic matter, the reason that small forests of recycled paper have been sacrificed to discussing its (hypothetical) properties, lies in the definition of nonbaryonic matter as capable of participating in the gravitational force while steadfastly remaining aloof from the other three forces at work in the universe. Thus, nonbaryonic matter does (by definition) just what the theoreticians want: It provides the gravitational force implied both by their theories and by the observations that we described in Chapter 11, while simultaneously avoiding the problem of interacting with familiar particles through other types of forces. Such interactions would have caused us to discover nonbaryonic matter by now. The fact that this discovery has not occurred demonstrates that nonbaryonic matter deserves its (etymologically distorted) name.

We have seen that neutrinos may be the nonbaryonic dark matter—if they have nonzero mass. Unfortunately for neutrino enthusiasts, the experimental upper limit on the mass of neutrinos has steadily decreased, and today it stands almost at the point that we can rule out neutrinos as the dark matter. If this turns out to be so, we shall be left with what some regard as an absurdity, but others see as the greatest challenge of the millennium in physics: Which of the hypothetical particles actually forms the dark matter that dominates the universe?

From a practical point of view, this question reduces to another: How can we find *any* of the hypothesized forms of nonbaryonic dark matter? If the answer to this question were easy, it would be wrong. Nonbaryonic matter, dark or not, by definition refuses to interact with our familiar, baryonic forms through anything other than gravitation and weak forces. The difficulty of detecting neutrinos testifies to how little interaction the latter represents. But that is what nature gives us, so anyone who searches for nonbaryonic dark matter must be prepared to search long and hard.

Today a number of dark-matter searches are under

way, funded by sympathetic governments in the United States and Europe. (With the Cold War over, we can no longer tell jokes about obtaining military funding for a "neutrino ray" or a "dark-matter bomb"—jokes whose punch lay in the fact that streams of neutrinos and nonbaryonic dark matter are something we all can and do face with reckless abandon every day.) Like the quest to find neutrinos, all these searches require a huge amount of shielding to eliminate other, baryonic (though relatively obscure) types of particles, which are less penetrating because they are far more highly interacting than the dark-matter particles we seek. Without a large amount of shielding, these less interesting, already known types of particles would overwhelm the sensitive detectors that scientists emplace to record (so they hope) their prey.

The leading locales where physicists today look for dark matter include experiments underneath the Oroville Dam in California, inside the Mont Blanc automobile tunnel on the Franco-Italian border, below the peaks of the Caucasus Mountains in Russia, and beneath the Gran Sasso, the highest of the Apennine Mountains in Italy. Each of these experiments tends to concentrate on a particular dark-matter candidate. For example, the Oroville Dam project aims to detect wimps (weakly interacting massive particles), whereas the Mont Blanc detector searches for magnetic monopoles. If one of the dark-matter candidates stood head and shoulders above the rest, we might well stop to focus on the particular experiment designed to find that type of particle. Instead, given the profusion of candidates, we may more appropriately admire the energy that physicists, both theoretical and experimental, can bring to bear on a modest problem such as the search for most of the universe. While they work on finding the dark matter, let us see how the discovery of dark matter might help us explain the structure of the universe.

THIRTEEN

World Enough and Time

Compared with the unknown future of universal expansion; the discovery of the cosmic background and its deviations from perfect smoothness; the inflationary theory, which posits that a tiny bubble of space took only 10^{-30} second to become far larger than the visible universe; and the dark matter, whose existence seems confirmed while its nature remains a mystery, what cosmological problem could be more difficult to understand? Just this: How did galaxies begin to form?

Nothing about the universe appears more evident than its clumpiness. We humans pass at will over the surface of a large clump of earth, which itself is dwarfed by comparison with the star that it orbits. This rather middle-class star, joined by its family of planets, their satellites, asteroids, comets, and meteoroids, moves as one of several hundred billion stars around the center of a huge spiral galaxy that we call the Milky Way. And strewn through space we find billions of similar galaxies, each of them a clump of matter divided into smaller ones and their attendants. The galaxies themselves typically cluster together; and on even larger scales of distance, we find that these clusters and superclusters of galaxies are arranged in bubblelike formations whose walls surround enormous voids in space. As astronomers have succeeded in studying the structure of the universe at successively greater distances, they have found a hierarchy of clustering—patterns that testify to long-range forces at work for billions of years.

No one is sure about the exact age of the universe, and recent evidence has shaken the previous view that the big bang occurred approximately 15 billion years ago. Strangely enough, according to our best theories, even 15 billion years could barely provide "world enough and time" (as Andrew Marvell put it) to allow the formation of the immense structures observed throughout the cosmos today. Recent estimates that only 8 to 13 billion years have passed since the big bang make the problem of explaining universal structure even more acute.

At this time, astronomers cannot tell whether the hierarchy of structure in the universe was established from the bottom up or the top down. In other words, we don't know whether the smaller clumps formed first, and only later assembled themselves into larger ones, or whether the largest ordering of matter, such as the soap-bubble patterns in the distribution of galaxies, formed first and later subdivided into smaller galactic units. However, because galaxies are the basic visible unit of the larger universe, we will first examine the possible mechanisms that shaped the galaxies themselves before attempting to determine whether galaxy formation preceded or superseded the formation of larger structures.

Galaxies exhibit great variation, though certain clear patterns are evident. The largest galaxies are either spirals or ellipticals, each containing hundreds of billions of stars and spanning at least a hundred thousand light years. Such enormous galaxies can be seen as faint smudges of light out to distances of several billion light years. Smaller galaxies, likewise spiral or elliptical, include only a few billion stars, or in some cases just a few hundred million stars; a small minority of galaxies of this size are "irregulars." No giant irregular galaxies have been found; the largest ones known do not exceed the sizes and masses of the Milky Way's two irregular satellite galaxies, the Magellanic Clouds.

In the neighborhood of the Milky Way, astronomers have discovered several dozen still smaller "dwarf galaxies," each with only a few tens of millions of stars. Since our own neighborhood is believed to represent a typical sample of the cosmos, astronomers readily conclude that dwarf galaxies exist throughout the universe. Their numbers probably exceed those of the more familiar, more striking "normal" galaxies, but their total stellar content falls far short of the amount of matter in the larger galaxies.

Giant, normal, or dwarf; spiral, elliptical, or irregular— all of these types of galaxies formed because of gravity. We know no other type of force that could have caused matter spread over distances the size of a galaxy to coalesce. Indeed,

The Visible Hand of Gravity

the matter that formed a galaxy must have originally spread through a larger-than-galactic volume, since every galaxy represents a clump of denser-than-average matter—a protogalaxy—that must have somehow drawn itself together. On rare occasion, astronomers have toyed with theories of galaxy formation using magnetic fields and electromagnetic forces, but these theories just don't work. In almost all circumstances, magnetic fields tend to inhibit gas clouds from contracting, rather than encouraging it.

The dominant role of gravity may seem surprising in light of the fact that gravity is by far the weakest of the four basic types of forces—weaker than electromagnetism, and much weaker than the so-called strong and weak forces that govern the interactions of elementary particles. Gravity owes its importance to two key attributes. First, gravity is a long-range force, like electromagnetism and unlike the strong and weak forces. Second, gravity always attracts. There is no antigravity, no negative mass that could cancel the familiar attractive effects exerted by all particles with nonzero, positive mass. Greater assemblages of particles always exert greater amounts of gravitational force. This fact explains why an initially modest clump of matter can attract more particles, and why the clump will tend to contract to a smaller size. Any protogalaxy will pass quite naturally, because of gravity, from a relatively ill-defined, spread-out clump of denser-than-average matter to become a smaller, denser object, a nearly formed galaxy.

When we ask, How did galactic structures begin to form? we should not lose sight of the big picture, which must include dark matter. Dark matter rules the cosmos, providing the bulk of the mass in the universe and therefore producing most of the gravitational forces that cause galaxies to form. We cannot expect to make ordinary (that is, non-dark) matter clump together without clumping dark matter as well. We might clump nondark matter *more* than dark matter, but the tendency to clump will appear in both. Otherwise the luminous tail would be wagging the dark-matter dog, which gravity forbids. Galaxies are places where all forms of matter have become far more concentrated than average, and of this matter, at least 90 percent, and quite possibly 98 percent, consists of dark matter. When cosmologists consider how matter clumped together, the fundamental question is thus, How did dark matter ever form clumps?

Therefore, in order to create theories that can explain the formation of galaxies, cosmologists must speculate about the long-vanished dynamical history of unseen matter of unknown form. It takes a peculiar bent of mind to turn one's fancy in this direction, but speculations, backed up by complex calculations, have not been lacking. By comparing the results of the calculations—that is, the predictions they make about the types of clumps of matter that gravity formed—with the galaxies we see, which are the ordinary-matter tip of the total (mainly dark-matter) iceberg, cosmologists can hope to reject some models and pass others on to the next stage of detailed investigation.

The Time Factor

All astronomers (with only fringe exceptions) agree that gravity must have made the galaxies. The problem is *time,* or rather the lack of time. Gravity works on marvelously large scales of distance, but it operates with ponderous slowness. Yet estimates of the ages of stars in our own Milky Way and in nearby galaxies suggest that many of them are 10 billion years old, and perhaps even older. Early in 1995 astronomers at the University of California, Berkeley, announced their discovery of a galaxy whose distance from us is so great that its light has traveled for four-fifths of the time since the big bang. This means that the galaxy cannot be more than about 2 billion years old as we see it. But this galaxy contains a fair proportion of reddish stars—which implies that the galaxy as we see it was not born yesterday.

In short, even allowing for uncertainties in dating the big bang, and putting aside the possible contradictions between the ages of galaxies and the age of the universe, it seems evident that galaxies existed as well-defined entities no more than 1 or 2 billion years after the big bang. And therein lies a paradox. To most members of the public, 1 or 2 billion years would appear time enough to do anything. But when cosmologists enter their data and theories into computers in order to come up with models that can build a galaxy in a couple of billion years, they repeatedly fail.

Though recent advances have brought us glimpses of very young galaxies, we have no views of galaxies during the actual formation process. The closest thing to a galaxy in formation may be the quasars, but these are almost certainly highly unusual galaxies seen early on. Views of protogalaxies may never be obtainable in visible light, since a protogalaxy has not yet formed stars and does not emit light at the wavelengths and frequencies that are familiar to astronomers.

Something younger than the most distant galaxies we can see, even younger than the quasars, is required if we are to understand the formation of structure in the universe.

How can we look backward in time, past the era when galaxies were young, to study the period when galaxies began to form? The oldest known relic in the universe is the cosmic background radiation, and concealed within it astronomers have a pearl of great prize. This radiation, produced during the first minutes after the big bang and free to travel without hindrance through space since the first million years or so, reaches us from distances of 8 to 13 billion light years. If we can decipher the imprint that gravity left on this radiation, we can thereby read the story of galactic history.

Seeds of the Galaxies

As Albert Einstein first showed, gravity affects photons, even though they have zero mass. Gravitational forces not only make photons deviate from straight-line trajectories through the bending of space but also affect the energy that each photon carries. For example, each photon escaping from the sun each lose a tiny fraction of its energy in fighting its way outward. The photons never cease to travel at the speed of light, but they do increase their wavelengths and decrease their frequencies (and energies) by about one two-thousandth of a percent as they leave the sun.

When gravity grows stronger, this effect becomes more pronounced. If the sun contracted its radius from 700,000 kilometers to a mere 3 kilometers, the gravitational force at its surface would increase to the point that any photon attempting to escape from its surface would lose all its

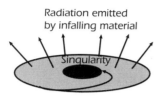

Radiation emitted
by infalling material

Singularity

Accretion disk
of infalling material

energy—and not escape at all! The sun would then be a black hole, an object whose gravity prevents all photons (and all other types of particles) from leaving. Black holes were once thought to be only theoretical constructs, amusing in their own mathematical sphere but not part of the real universe. Today, though, astronomers take them quite seriously—despite only circumstantial evidence (one could hardly expect direct observation) for the existence of any black hole. Many strange sources of intense radio emission are believed to arise from matter spiraling into black holes.

When we examine the effects of gravity on the photons in the cosmic background radiation, the situation becomes simple in concept, though amazingly complex in detail. Any greater-than-average concentration of matter at the time that the cosmic background radiation was produced will affect the background photons, causing the number of photons of a given energy in a given volume to differ from the average. A lower-than-average density will produce a similar effect, but in the opposite direction. The greater the deviations from the average density along a particular line of sight outward into the universe, the more marked will be the deviation from the average that we observe in the number of photons in the cosmic background radiation of a particular wavelength and frequency.

Therefore, in principle, all that we must do to determine the concentrations of matter at the time that the cosmic background radiation was produced is to create a high-precision map of the sky, showing the intensity of photons in the cosmic background radiation. This map would reveal how the radiation was affected by concentrations of matter in the early universe and should reveal the deviations from perfect smoothness—if any existed—during the first billion years after the big bang.

Before plunging into this effort, however, we should note one difficulty in judging what these irregularities can tell us. It may turn out that most of the irregularities in the cosmic background radiation arise not from perturbations in the density of matter, which are good for forming galaxies, but from gravity waves themselves, which are not.

Gravity waves, also called gravitational radiation, travel at the speed of light, but they differ from all other types of radiation in the universe. They arise whenever matter moves rapidly, and they have an odd effect on objects as they pass. Instead of moving the object, they distort its shape, elongating it in one direction perpendicular to the waves while shrinking it the other direction across the propagation direction. Then the elongation and shrinkage directions reverse, and reverse again, as the waves pass.

The early universe, which was filled with matter moving at enormous speeds, must have produced significant amounts of gravitational radiation. These gravity waves could have been responsible for at least part of the irregularities in the cosmic microwave background; just how much remains an open question. Because these irregularities do not represent perturbations in density, they are of no use in explaining how galaxies began to form. In the cosmic background radiation, at least part of the deviations from perfect smoothness—perhaps the majority—could arise from gravity waves. This would leave even fewer of the already modest density irregularities as the "seeds" of galaxies. But for now, let us put this problem out of mind and examine what the irregularities in the cosmic background radiation can tell us—if they arise from the perturbations in density that could form galaxies.

COBE: The Cosmic Background Explorer

On April 24, 1992, cosmology burst into the lead position (to be sure, on a slow day for news) on television news programs and the front pages of the world's newspapers. On that day, readers and viewers learned that a satellite called COBE had found evidence of "ripples" in the cosmic background radiation—ripples that provided the oldest evidence for the variations in density from place to place in the universe that could later have grown into galaxies (see Figure 10). Doubtless the general public understood little of this; what they did understand was that scientists had had some sort of experience that was "like looking at God."

Scientists who talk to the public, aware that the public can mute its excitement about science quite easily, often attempt to convey their emotions by using "God" as a metaphor for what most of us would call "nature." This does not mean that scientists are generally irreligious. Like the general public, scientists embrace philosophical ideas that range from extreme atheism through modest agnosticism to all forms of conventional and unconventional religious beliefs. Many scientists share the usage made famous (at least among scientists) by Einstein, who expressed his distrust of the quantum theory by insisting that "God does not play dice" and his belief that science could unravel nature's secrets by stating that "God is subtle, but not malicious." Pressed for a statement of his religious beliefs, Einstein replied that "I believe in Spinoza's God who reveals himself in the orderly harmony of what exists, not in a God who concerns himself with fates and actions of human beings." This could well sum up the attitude of many, but by no means all, of today's cosmologists.

When it comes to popularizing science, however, cosmologists have learned that references to God provoke immediate and widespread interest. The announcement of what COBE had discovered provides a fine example of this phenomenon. Astronomers who noted that the results are "highly significant" went unquoted in the press, while those who referred to irregularities in the cosmic background radiation as the "Holy Grail" of cosmology, or said that observing these irregularities was "like finding the handwriting of God" received worldwide attention.

As is the case with all campaign rhetoric, the press elides the caveats and dispenses with the details. At the key press conference describing the COBE results, George Smoot, the leader of the team that discovered the irregularities in the cosmic background, carefully stated that "*if* you're religious, it's *like* looking at God" (italics added). This quote was soon shortened to "Like looking at God," and then, on the cover of *Maclean's*, Canada's leading newsweekly, to "Looking at God." Smoot's co-workers at the Lawrence Berkeley Laboratory reacted to the eagerness with which the press took up this quote by irreverently mounting in the corridor a copy of the COBE map with the caption "Behold the Face of GOD." Since the map that was published around the world does not actually show the irregularities that

COBE found—these must be dug out from the data through laborious statistical analysis—the quote has an apt ring for those who ponder the fate of humanity.

But enough of theology in the service of mankind. What were the implications of the COBE discoveries that received worldwide, if brief, attention and admiration? The Cosmic Background Explorer satellite had been designed and launched (in November 1989) to make a long-term study of the cosmic background radiation (see Chapter 6). Ever since this radiation was first detected in 1964, cosmologists have known that the details observed in the radiation would carry information about the state of the universe at the time that the radiation last interacted with matter, some 300,000 years after the big bang.

One of COBE's key instruments was the Differential Microwave Radiometer, or DMR, designed to compare the amounts of radiation arriving from different parts of the sky. Although the longest wavelengths and lowest frequencies of the cosmic background radiation can penetrate our atmosphere, most of the radiation cannot. We therefore must send instruments into orbit to make a long-term study of this cosmic relic.

In January 1990 COBE completed its measurements of the spectrum of the cosmic background radiation with an accuracy far greater than those available from previous observations. The spectrum that COBE measured matched perfectly with the prediction made by the big-bang model of the universe (see page 74). Today only a handful of cosmologists continue to suggest alternative models of the universe in which no big bang has occurred. In these models, the cosmic background radiation must be explained in a fairly ad-hoc way, as a much more local phenomenon that appears evenly distributed on the sky. Most astronomers consider this impossible, because the cosmic background radiation has near-perfect smoothness all the way around the sky.

Had COBE done no more than to make stunningly precise measurements of the spectrum of the cosmic background radiation, its team of scientists and engineers, led by NASA's John Mather, might well have pronounced themselves satisfied. But COBE had been designed to do something still more spectacular: not simply to measure the spectrum of the cosmic background radiation but to make a detailed set of observations that might find tiny deviations in the amount of this radiation arriving from different directions in the universe.

A nearly smooth spectrum of radiation was essential to proving the big bang, but a *completely* smooth early universe would be unbearable to theorists, because it would mean there were no primordial concentrations of matter to get the galaxies and larger structures off the ground, so to speak. The theorists hope for—demanded—*some* irregularities. All previous observations of the cosmic background radiation had revealed no deviations from complete smoothness in the radiation. This result was immensely worrisome to cosmologists, since they were encountering great difficulties in forming galaxies on their computers.

To make its crucial observations of fluctuations in the cosmic background radiation, COBE's DMR instrument was designed to perform one crucial operation: mapping the sky in radiation whose frequencies lie close to those where the cosmic background radiation has its peak emission. Previous maps of the background radiation had insufficient sensitivity to reveal any deviations from the average—except for the changes produced by the Doppler effect arising from our galaxy's motion with respect to the local average.

The DMR instrument on COBE observed the cosmic background at three different frequencies, but it measured only the *difference* between the amounts of radiation received from two points on the sky 60 degrees apart. This makes good sense, because an instrument such as the DMR can measure these differences with far greater accuracy than it could derive them by measuring the actual amounts of radiation and later subtracting one amount from the other. COBE took measurements over a full three-year period, and every six months the DMR completed a map of the sky in all directions, with its ground controllers exerting great care to assure that COBE never pointed toward the earth or the sun, both of which were bright enough to put COBE out of business.

Each of the DMR's individual observations produced a "pixel," short for "picture element," spanning an area a bit less than 3 by 3 degrees in angular size. Nine square degrees amount to 40 times the portion of the sky that the full moon covers; since the entire sky includes something over 40,000 square degrees, more than 5,000 such pixels cover the entire sky. However, in order to make an accurate map from a measurement of the difference in the strength of the radiation, each pixel must be observed thousands of times.

During its first year of operation, the DMR experiment

collected nearly 200 million pixels, observing each area repeatedly in two entirely separate channels (to check on instrument noise) for each of three different frequencies of the background radiation. Finally, confident that DMR had found real irregularities in the radiation, and not just instrument noise, the DMR science team proceeded to the heart of their project—analyzing the pixels to see whether any patterns of irregularity could be found.

This was a monumental task. Two important patterns were known to exist (in addition to the unknown effects of gravity waves) that do *not* reveal irregularities in the early universe, and must be subtracted from the observations before the search for cosmic irregularities can proceed. One pattern is the emission of radiation from our own Milky Way galaxy, which adds to, and interferes with, the radiation from much farther away. The second pattern arises from the cosmologically modest motions of the solar system and the Milky Way in space. These motions produce a characteristic pattern in the cosmic background radiation that we detect: The radiation has a slightly higher temperature in the direction toward which the Milky Way is moving, and a slightly lower temperature in the opposite direction.

Achieving a sensitivity not previously available, the DMR could detect deviations of approximately one part in a hundred thousand from complete smoothness in the background radiation. Nearly all cosmological theorists agreed in advance (but would they have kept their promise?) that *if* the DMR failed to detect any irregularities

Not Everything Is the Background Radiation

in the temperature of the background radiation, they would have to abandon their most-beloved models (typically one per theorist). Failure to find fluctuations of even one part in a hundred thousand, dating from a few hundred thousand years after the big bang, would mean that no model yet imagined could explain how the universe managed to produce complex structures high in density contrast a few *billion* years later.

Armed with years of experience, the DMR team allowed for the effects of the Milky Way and our galaxy's individual motion, and seized their prize. "We have a quadrupole," Smoot reported in Washington, using language that describes deviations from smoothness. The irregularities encapsulated in this word are the oldest and largest structures yet seen in the universe. Thanks to the expansion of the universe, the *smallest* of these irregularities now spans a region larger than the Great Wall of Galaxies. But what COBE found are not structures in the traditional sense. Instead, they are regions whose temperatures at the era of decoupling differed by about three parts in a hundred thousand from the average.

Nevertheless, the irregularities that the DMR experiment has found are immensely significant. As the first deviations from smoothness seen in the glow from the early universe, they provide crucial support for all theories that explain the formation of galaxy clusters by gravity. Given its level of sensitivity, had COBE not found the deviations from smoothness that it did, this chapter might have been called something like "Cosmologists Seek Bold New Ideas."

The all-sky map produced by the DMR on the COBE

satellite appeared in a thousand newspapers, often so poorly explained that readers were left to wonder why the universe has the shape of a watermelon. (In fact, the melon shape arose simply from the way that astronomers plotted the data that the DMR obtained by observing all around the sky.) The mottled pattern show different levels of intensity of the radiation: the long-sought deviations from average in the cosmic background radiation. Ready for the moment, the scientists receiving the DMR data performed a host of statistical analyses on the data represented in this map. To this day, arguments continue as to exactly what the mottling reveals, though certain conclusions appear certain.

A crucial result from the DMR lies in the ratios of the numbers of irregularities with different sizes. These ratios reveal which types of cosmological models can furnish a viable explanation of galaxy formation.

What Do the Wrinkles in Time Tell Us?

Again we confront the essence of science: Create models, if you can, that attempt to explain what we see, and then toss out the models that cannot conform to additional observation. For example, the DMR results ruled out an entire class of models, those based on "cosmic defects." A cosmic defect is a region of immensely high density that might have provided one of the seeds of galaxy formation through gravity. Some of the best-known cosmic defects are cosmic strings, particles hypothesized to be nearly infinitesimally thin but so massive that every millimeter would weigh more than Mount Everest. The cosmic-defect models predict far more small-scale than large-scale irregularities. But COBE found approximately the same number of irregularities of all sizes within the range that it measured. For now, cosmic-defect models seem ready for the wastebasket of ideas that look fine on paper but do not describe the actual universe.

But perhaps we have not heard the last from the theorists who originated these models, and indeed (to say it once more), only through continual changes to these models, and repeated testing of the new models, can we hope to achieve the final acceptance or rejection that we (collectively) seek. If we agree that the DMR observations push the cosmic-defect hypothesis to a lower position in the hierarchy of competing models, which ones, we may ask, have risen higher from the COBE results? Most notably, the inflationary model, which predicts a spectrum of irregularities similar in their deviation from complete smoothness at different angular scales to the one that the DMR observed.

When the inflationary theory was developed during the early 1980s, few of its creators noticed that in its simplest form the model makes a prediction about any irregularities that arose in the primordial universe. According to inflation, we should find irregularities of all different sizes in about the same numbers; the super-rapid expansion of the universe would have "inflated" all the original irregularities in the same way. The prediction of equal numbers of irregularities of different sizes was a by-product of the basic inflationary model, made at a time when no one expected irregularities to be measured soon. For this reason, the fact that the DMR experiment found what the inflationary model predicted years earlier was far more impressive than it would be if the model had been conceived to explain the ratios of observed irregularities. Paul Steinhardt of the University of Pennsylvania, one of the originators of inflation, reacted to the DMR results by realizing, "This makes [the inflationary model] real."

"Real" means different things to different people. Most cosmologists, having rejoiced that they finally had actual data on the fluctuations in the early universe to argue about, quickly began to ask what further observations could be obtained to discriminate among models. In seeking to test the inflationary model, or any other model purporting to describe the early universe, cosmologists now have two needs. First, they want data that will either confirm or disprove the irregularities already observed by the DMR experiment. Second, they want to find and to measure irregularities in the cosmic background radiation at smaller angular scales than the DMR can study—angular scales that correspond to the formation of galaxies and galaxy clusters. Fortunately, cosmologists should not have to wait long. A nicely interlocking combination of experimental evidence is scheduled to emerge within the next few years.

Scientists continue to analyze the data from the COBE satellite, seeking to improve the conclusions announced in 1992 and refined through two more years of COBE observations. Three other experiments, similar to COBE in concept but different in key details, are now under way at different places on Earth to assist this effort.

First among these is a series of balloon flights to carry detectors 100,000 to 120,000 feet above Earth, to altitudes

Further Observations of the Cosmic Background Radiation

which most of the cosmic background radiation reaches before our atmosphere blocks it. The second type of observations are those made from the South Pole, at an altitude of 9,200 feet, where the cold, thin air is remarkably free of the water vapor that absorbs most of the cosmic background radiation. In the third approach, conventional radio observatories, such as the Owens Valley Radio Observatory in California, study the cosmic background radiation in fine detail by linking several dishes together to form an "interferometer." Although only a small fraction of the cosmic background radiation can penetrate to the 4,000-foot elevation of the Owens Valley Radio Observatory and its fellows around the world, that fraction can be analyzed far more efficiently than can the radiation measured by any balloon-borne or South-Pole detector.

All three types of experiments have already gathered data, and all have received a great psychological impetus from COBE's results. Greater attention now focuses on the balloon-borne and South Pole experiments, because they can detect the most important frequencies in the cosmic background radiation. With newly improved instruments, these experiments can search for irregularities with a sensitivity just about equal to that of the DMR experiment, though they do not survey the entire sky. But these experiments do examine small patches of the sky with an angular resolution much finer than that available with the DMR instrument on COBE. The DMR has an angular resolution of 7 degrees, which means that it cannot distinguish objects or patches of sky that are not at least 7 degrees apart. In contrast, the South Pole instrument can survey the sky with a resolution of .75

degrees, nearly 10 times better, and the balloon-borne detectors examine the cosmic background radiation on still finer scales, with an angular resolution of just half a degree—the angular size of the sun or moon. This resolution falls far short of that which can be achieved with ground-based radio telescopes, which attain an angular resolution of about one-tenth of a degree. However, unlike the balloon and South Pole detectors, radio telescopes on earth's surface can observe only frequencies far from the peak frequency of the cosmic background radiation.

Nevertheless, because these telescopes can achieve high angular resolution, they may prove equally decisive in interpreting the details of the cosmic background radiation. These details include the ratios of the sizes of the irregularities of different angular sizes, and the correlations between irregularities in one part of the sky and those in neighboring parts. They offer cosmologists the chance to determine not only the importance of gravity waves in the early universe, but also whether or not the inflationary model, which makes specific predictions about these details, is correct. Once again, the reader would do well to stay tuned for further developments, because the answers are not yet at hand. However, before the millennium arrives, cosmic-background astronomers hope to find irregularities as "small" as those that became clusters of galaxies, and to use their detailed observations to reject or accept competing models of the early universe.

FOURTEEN

Hot Dark Matter, Cold Dark Matter, What's the Matter?

Dark matter, with its dominating mass and gravitation, provides cosmologists with their current best hope to determine the density, age, and future of the universe, to validate (or reject) the inflationary version of the big bang, and to explain the ever larger structures astronomers see as they increase the distance scales of their observations. But gravity alone is not enough to explain the evolution and structure of the universe, as we have seen. Cosmologists must rely on fluctuations from perfect smoothness, perhaps left behind by the inflationary era, to help galaxies begin to coalesce. If the right sort of fluctuations existed then, gravity would have increased the contrast in density over time, until clouds of matter reached densities that would allow stars to began to form within them.

So far, so good. But cosmologists are nowhere near agreement on what makes up the dark matter in the universe. Different models produce different kinds of "galaxies," depending upon what type of dark matter is incorporated into the model and how that type affects the type of clumps that the matter will form.

Cosmologists' basic strategy for investigating these questions has been to choose a model for the dark matter and the initial types of perturbations or "seeds," and then to turn the computer loose to see what happens. After a while (far less of a while than was true with earlier generations of computers), they examine what the computer hath wrought (see Figure 9). If it looks like the real universe, that's good. If it doesn't, well, it's still pretty good if it has aesthetic or other appeals to the cosmologist's soul.

During the past decade, many proposals for just this sort of cosmic investigation (posed in terms somewhat more sober than the description in the preceding paragraph) have been submitted, funded, executed, and published. In order to grasp the broad outlines of what humans have achieved with these studies, we must examine the final (for us) and greatest classification of possible dark-matter types: hot dark matter and cold dark matter.

Hot Versus Cold

In cosmologists' language, the words, "hot" and "cold" correspond only vaguely to their familiar meanings. We all have learned that in a gas, each individual particle—for example, the helium atoms in a balloon—move and collide at random, free of any links to their particle neighbors like those that exist among the particles in a liquid or solid. Within a gas, temperature measures the average kinetic energy per particle. This kinetic energy of each particle varies in proportion to its mass times the square of its velocity. When physicists discuss the velocities of particles in a gas, they therefore fall naturally into the habit of specifying the temperature within the gas: Higher temperatures go with larger velocities, and vice versa.

Cosmologists carry this usage into their descriptions of the particles in the universe at the time that galaxies began to form. In this context, "hot" particles are those with velocities extremely close to the speed of light, while "cold" particles are those whose velocities fall far below the speed of light. (Since the temperature of matter in the universe declines as the universe expands, hot matter will eventually become cold matter if enough time passes. Indeed, what we detect today in the cosmic background radiation are "warm" particles, with speeds slightly less than the speed of light. Of course the era that counts for forming galaxies is not today but rather the time when the matter began to form clumps. During that era, shortly after the big bang, the difference in clumping behavior between the two classes of matter was crucial.)

The key difference between hot-dark-matter models and cold-dark-matter models is this: Because hot-dark-matter particles are moving far more rapidly than cold-dark-matter particles, hot dark matter quickly escapes from any small region of space. Hence any clumps that hot dark matter forms will be much larger than those that cold dark matter will make. Just how much larger depends on the masses of the hot-dark-matter particles. If, for example, the dark matter consists of neutrinos with masses equal to about

1/100,000 of an electron mass, then the smallest clumps that could have formed since the big bang have masses of about 10^{16} times the sun's mass. Since a large galaxy has a mass of "only" 10^{12} solar masses, we are discussing clumps with ten thousand times the mass of the Milky Way. Thus, hot dark matter—at least in the form of its prime candidate, the neutrino—will produce clumps whose masses resemble what we find in the largest clusters of galaxies.

If the real universe works this way, then some *other* process must have made the basic structure of the universe—galaxies—from subclumps that formed within the galaxy-cluster-sized clumps that hot dark matter produced. Furthermore, if this scenario contains the kernel of reality that we seek, we could say that the universe was formed in a top-down, as opposed to a bottom-up, fashion—with the largest structures forming before the smaller ones (see Diagram 8).

In contrast with hot dark matter, cold-dark-matter models all impose a bottom-up order to the formation process (see Diagram 9). Cold dark matter forms small clumps extremely well, because the matter does not tend to "run away" from the invisible gravitational hand that pulls it toward a denser-than-average perturbation. But cold dark matter fails to form clumps of sufficiently large size to satisfy all the requirements of a completely successful theory.

Because the particles in cold dark matter have no significant random motions, the precise mass of the hypothesized particles plays no role in the calculations: They are all sufficiently massive to be "cold." We therefore cannot hope to discriminate among cold-dark-matter models by the type of clumps they produce, for all results are essentially the same. We may regret this the more when we reflect upon the large number of cold-dark-matter candidates that currently exist in modern cosmology.

The clumps that cold-dark-matter models generate range from masses of 10^8 to 10^{15} solar masses (which typify galaxy clusters), but they form clumps preferentially at the low-mass end of the distribution. And here the difficulty arises: 10^8 solar masses amounts to a dwarf galaxy, whereas the giant spiral and elliptical galaxies that dominate our view of the cosmos average 10^{12} solar masses. One natural reaction to this result might be to conclude that the smaller objects produced by cold-dark-matter models have themselves clumped into larger ones, typified by giant galaxies.

Unfortunately, the cold-dark-matter models have another obstacle to overcome, one which arises from the inflationary theory of the big bang. When cosmologists assume, in accordance with the inflationary model, that the density of the universe equals critical density, their computer-generated universes produce far too many clumps of relatively small mass (10^8–10^{10} solar masses), and not enough clumps of large mass. Cold-dark-matter models do work well if the average density of matter amounts to only about 20 percent of the critical density. In this case, the problems of over-clumping at small scales does not become so severe that we cannot fit the model to the observations.

It probably did not escape the reader's notice that an average density of matter equal to only 20 percent of the critical density comes close to our observational data. These data suggest that the density of dark matter amounts to at least 5 to 10 times the density of visible matter. It is the inflationary model's successes that drive forward the quest to find a workable model with a density of matter *equal* to the critical density. If the density of dark matter falls well below the critical density, the cold-dark-matter models seem favored, but if we seek a viable model for the cosmos with a density close to the critical density, the hot-dark-matter models, partly by default, come to the fore.

Is the Dark Matter Made of Machos?

Some astronomers have investigated the possibility (first proposed by theorists) that *machos* (massive compact halo objects) might account for the dark matter in the universe. Machos fall into the general category of cold dark matter, but they are not elementary particles; instead, they are relatively large objects. Machos might be large rocks, asteroid-sized objects, or even massive planets something like Jupiter—all baryonic.

But machos might be stranger still: They might not consist of "ordinary" baryonic particles but may instead be made of nonbaryonic matter. Because astronomers detect machos solely by their gravitational effects, they cannot hope to determine what type of matter forms a macho without other ways to study them. But since machos produce little or no radiation on their own, these "other ways" seem likely to be lacking indefinitely. Machos might even turn out to be black holes. Although a black hole lets no matter escape from it—not even photons or other massless particles—gravity does "escape": A black hole affects the universe by the gravitational force that it produces.

It now appears that machos are spread throughout the galactic halo of the Milky Way and other galaxies like it. Astronomers' hopes for finding machos rest with another aspect of Einstein's theory of general relativity, which predicts that gravity bends light—or, if you prefer an alternative view of the same result, gravity bends space, so that light rays that would otherwise follow straight-line trajectories adopt curved paths in the presence of large amounts of gravitational force. This part of Einstein's general relativity first received direct observational confirmation during the total solar eclipse of 1919, and has been repeatedly verified since then.

A macho located almost directly between ourselves and a distant star will bend the light from that star. The macho will not only bend the light but will also tend to focus it. As a result of this "gravitational lens effect," when the macho passes almost directly between ourselves and the star, the star will appear to grow brighter for the few days when the macho lies most nearly along our direct line of sight to the star.

Could astronomers hope to detect such an event among the billions of stars subject to their gaze? And how could they distinguish it from the temporary brightening that variable stars undergo all by themselves? The answers lie in the details, and are basically: "Yes," and "With hard work." During the past few years, astronomers have found events that they assign to the macho deviation of starlight. An ongoing search for machos, which employs an otherwise unused telescope at Mount Stromlo Observatory in Australia, has detected 3 objects along the line of sight to the Large

Magellanic Cloud and several dozen in the central regions of the Milky Way.

This macho search is now starting to acquire a sufficient statistical base to draw significant conclusions. The survey can discover objects whose masses range from several times the mass of the sun down to about the mass of Mercury (1/10,000,000 of the sun's mass), by observing the gravitational lens effect that these objects produce on rays of starlight passing close by them. The preliminary results show that machos in this mass range may provide the cosmos with a density of matter as large as 10 percent of the critical density, but not more. In other words, if the preliminary conclusions are proven to be correct during the next few years, we shall know that machos do exist—in a form that remains to be determined. Machos may provide the bulk of the dark matter, but according to the present results, they cannot make the universe cease to expand.

Massive compact halo objects that consist of baryonic matter would also be unlikely to provide much of the dark matter, for the reasons we discussed in Chapter 10. Until this assessment changes, we must look to candidates other than machos to explain the nature of the dark matter.

Could the Dark Matter Be Both Hot and Cold?

One can easily see where this hot-versus-cold dilemma leads. A cosmological model in which the unseen mass that dominates the universe consists partly of hot dark matter and partly of cold dark matter yields results superior to those from either type of dark matter alone. What has Occam's razor to say to this?

The razor tells us that we shouldn't add another entity to the universe unless we must; and to many cosmologists, two types of dark matter is one type too many. An enthusiast might reply that since we can accept that the universe contains at least 5 to 10 times more dark matter than visible matter, dividing the dark matter into two types would still leave plenty of material for each type. The enthusiast might add that if theorists have produced several dozen possible candidates for the dark matter, each of them justifiable on some grounds, to accept two of them as real amounts not so much to a fudge factor as to a tremendous winnowing and rejection of hypotheses.

Particle physicists have gone one better. In 1993, responding to the perceived need for two types of dark matter, scientists at the Canadian Institute for Theoretical Astrophysics have hypothesized a new type of neutrino that would decay into hot-dark-matter *and* cold-dark matter particles! In this (completely hypothetical) process, the neutrinos imitate what a laser does. Inside a laser, one photon "stimulates" the emission of other identical photons, which in turn stimulate still more emission. The result is a stream of coherent light—a beam of photons with identical wavelengths and frequencies. When the hypothetical neutrinos each decay into two particles, one of them (the cold-dark-matter type) stimulates the production of more neutrinos, which likewise each decay into two particles, of which the more massive one stimulates still more neutrino production.

This would be highly significant if true—a remark that can be made about most of the current work on dark matter. Two-fold dark matter may turn out to describe the real universe, and (who knows?) may be only a dual aspect of a single type of particle. To quote a favorite line from many grant proposals, further research will be needed. But for now, since we are laying on the table all the cards that seem appropriate, let me mention one final contribution to the hypothetical happiness achieved to date.

A Little Bias

Much of what I have described about forming galaxies on the computer for comparison with the real universe involves a factor that theorists call "biasing," but which I would characterize as a fudge factor. Biasing refers to the fact that the distribution of galaxies may not trace the distribution of dark matter. Since dark matter rules the universe by gravitation, its clumping is what the computer follows. Ordinary visible matter merely tags along while the 5-to-100-times-denser dark matter either clumps or does not. But in the real universe, galaxies *may* appear only where the dark matter—and the accompanying ordinary matter—becomes a particularly dense clump. In other words, the structure that we are able to observe may be the densest clumps, and many other not-quite-so-dense ones may be sprinkled throughout the universe. Indeed, this seems quite reasonable if we assume that some minimum density of ordinary matter must be reached to "trigger" the onset of star formation, the process that made stars shine.

If this is true, then a cluster of galaxies may represent a region in which many peaks in the density distribution poke above the minimum level needed for star formation. The actual clumpiness of the universe—the clumpiness of the dark matter—might be considerably less pronounced than the clumpiness that we see, because the biasing introduced by the minimum-density requirement for star formation would leave all the lesser clumps invisible.

Biasing plays a key role in cold-dark-matter models, which tend to produce too many small clumps, spaced too closely to mimic the real universe. With biasing, most of these clumps would never form stars or galaxies. With biasing, we "lose" many of the smaller clumps, and gain a result much more like what we see. We may fairly say that *without* biasing, current cold-dark-matter models are dead.

Of all the fudge factors that we have met in our cosmological wanderings, biasing seems one of the best. Cold dark matter and hot dark matter each have their advocates; and in persuasiveness, their arguments are fairly evenly balanced. Nearly every cosmologist will agree that no model works in its simplest form; it is up to us (speaking for the entire human attack on understanding the cosmos) to determine how complex the models must be in order to explain what we see, and what reasons exist for accepting or rejecting these models.

What hopes do we have to resolve the differences between the hot-dark-matter and cold-dark-matter models of the universe, to measure the density and, thereby, determine the actual age of the cosmos, and to perceive how the largest structures came into existence? Improved computers will help, but what we really need are better observations, either of the distribution of matter in space, or of the primordial seeds that led to galaxy formation, or of the actual dark-matter particles about which we can now only speculate.

Observations of the actual distribution of matter will require decades of careful and tedious work, but will surely provide us with an improved knowledge of the structure and density of the universe. The second quest, for more information about the seeds of galaxies, will almost certainly become available within a couple of years. It will allow us to exclude some models of galaxy formation and pass others along to the next level of investigation. The third search, for dark matter itself, might succeed, even next year, in revealing the major component of the universe. But we might conceivably continue for a long time without being able to do more than to exclude certain hypothetical dark-matter particles while leaving us with an oversupply of candidate types.

As I reach the end of this survey, I am torn between emphasizing how much we have learned and stressing how little we know of the cosmic processes that took us from the big bang to our present situation. Margaret Geller likes to compare our difficulties in understanding the history of the universe to a moviegoer who sees a few frames at the start of the film, then falls asleep until the movie has nearly ended, and quite understandably must employ great mental exertion even to guess at what happened in between. In this analogy, the early frames correspond to observations of the cosmic background radiation, which show tiny deviations from an absolutely smooth universe, and the late frames represent our current view of the universe. Even though we look far back in time when we observe galaxy clusters billions of light years away (see Figure 12), we cannot begin to see the crucial missing frames, which span the period between a million years after the big bang, when the background radiation began to travel freely, and a couple of billion years later, when quasars and galaxies had already formed in immense numbers.

During the next decade, better telescopes, along with greatly improved infrared detectors, offer the hope of finding some of the missing frames, on which the Doppler effect has made the picture—radiation from young galaxies and still-forming galaxies appear primarily in the infrared. Early in the next millennium, we may therefore have a much better understanding of how galaxies formed. If astronomers also determine the true value of the Hubble constant, as seems likely, and can resolve the question of how old are the oldest stars, then we can congratulate them on a fine set of accomplishments. But it would be foolish to imagine that this will settle today's burning cosmological issues, which include the questions of how galaxies formed, what most of the universe is made of, and what future lies in store for the universe. And we may confidently predict that by the time this book lies hidden below piles of more recent contributions to scholarship, great new cosmological questions will have appeared, crying out for answers as lustily as any do now.

INDEX